Career Options in the Life Sciences: Guide Research for Your Path Beyond Academia

AF806255

Stefano Gaburro • Eric Rieux

Career Options in the Life Sciences: Guide Research for Your Path Beyond Academia

Springer

Stefano Gaburro
Berlin, Berlin, Germany

Eric Rieux
Choisy, France

ISBN 978-3-031-73564-6 ISBN 978-3-031-73565-3 (eBook)
https://doi.org/10.1007/978-3-031-73565-3

© The Editor(s) (if applicable) and The Author(s), under exclusive license to Springer Nature Switzerland AG 2024, corrected publication 2024
This work is subject to copyright. All rights are solely and exclusively licensed by the Publisher, whether the whole or part of the material is concerned, specifically the rights of translation, reprinting, reuse of illustrations, recitation, broadcasting, reproduction on microfilms or in any other physical way, and transmission or information storage and retrieval, electronic adaptation, computer software, or by similar or dissimilar methodology now known or hereafter developed.
The use of general descriptive names, registered names, trademarks, service marks, etc. in this publication does not imply, even in the absence of a specific statement, that such names are exempt from the relevant protective laws and regulations and therefore free for general use.
The publisher, the authors and the editors are safe to assume that the advice and information in this book are believed to be true and accurate at the date of publication. Neither the publisher nor the authors or the editors give a warranty, expressed or implied, with respect to the material contained herein or for any errors or omissions that may have been made. The publisher remains neutral with regard to jurisdictional claims in published maps and institutional affiliations.

This Springer imprint is published by the registered company Springer Nature Switzerland AG
The registered company address is: Gewerbestrasse 11, 6330 Cham, Switzerland

If disposing of this product, please recycle the paper.

Preface

From Eric:

When Stefano invited me to join in writing this book, I felt both excited and honored. Stefano and I have a long history of collaboration, friendship, and respect for each other. While I helped Stefano start his career outside Academia, Stefano helped me become a better leader and manager thanks to his stimulating questions and healthy challenges. Over the last 20 years, I had the opportunity to directly recruit several science graduates, and help many more start and then grow their careers in the private Industry. Seeing others build fulfilling careers has been my greatest reward. I hope this book will guide you on your personal journey.

From Stefano:

I have been working in academic settings for 12 years. Due to health reasons, I was forced to leave academia in 2012 after winning an EU grant on a cure to Fragile-X and getting a tenure position at Charite in Berlin. Fortunately, I had the chance to meet Eric along my path. Eric saw in me the "sales" person I was unaware of as a scientist who always focused on molecular pathways in neuroscience specific to behavioral translation. Fast forward, I quickly realized the importance of developing my soft skills, and I have always looked for different fields "out of science" to bring them back into my date-to-date business and apply them to my needs. I have helped 125+ researchers voluntarily move out of academia and succeed in their positions, proving that a career transition is possible and rewarding with the proper guidance and practical steps. This reassurance should give you a sense of security and preparedness as you embark on your transition journey.

Berlin, Germany Stefano Gaburro
Choisy, France Eric Rieux

Acknowledgments

From Eric:

While I have written a lot of content for various types of corporate reports, marketing and sales documents, training materials for staff, technical sections of scientific articles, and on the fun personal side, short stories and theatre plays, co-writing this short, structured book has been a new and rewarding experience. It is never too late to learn new skills! Thank you Stefano for the partnership on this project.

To my friends and colleagues whom I interviewed directly for the purpose of this book, Barbara, Eric, Jean-Pierre, Sébastien, and Yanglei, my most sincere thanks for your time and valuable insights; your life stories are so inspiring. To my dear spouse Karen, herself a Science PhD and accomplished published writer, thank you for your lifetime support and encouragement for this book.To my friend Gary, editor of the monthly WHO newsletter, "thanks mate" for your insights and many tips. I would like to finish by thanking all my friends, colleagues, clients, and partners in the Biomedical field for the years of collaboration, shared learning, and opportunities. I am grateful for the rewarding career I have had; you are all part of my life story.

From Stefano:

I started this project inspired by the 125+ stories I have encountered. I thought that sharing my current knowledge in a written form (book) would encourage others in the STEM areas. Thanks, Eric, for participating in this.

I wish to thank my wonderful wife, Katharina, for the past, present, and future support you have given me and my family and friends, who push me to grow daily.

I also thank Michele, Torben, Frederick, Elena, Cindy, Mushtaq, Diana, and Bruno for sharing their success and inspiring stories for the success of the book.

Ultimately, I would like to thank all my colleagues at Tecniplast and the clients I met along the way for sharing their insights and for the great learning opportunities I had.

Introduction

In our life science education, many of us, myself included, underwent classical life science academic training, preparing to perform research at the university. However, according to an anonymous survey we recently conducted, only 20–30% of those students who enter life science education end up in permanent academic positions (see Fig. 1). The rest of us are leaving academia because of a lack of permanent positions (lectured, professor, or tenure track) [1, 2], the underpaid PhD/Postdoc offerings, and the publish or perish system that, in turn, might generate a lack of funding or eventually a toxic Principal Investigator (PI) to student relationship.

The graph illustrates the diverse career trajectories of new PhD and postdoctoral recipients, with a notable proportion entering the academic sector, albeit not exclusively in tenure-track positions. The representation of industry, government, and non-profit sectors underscores the wide range of opportunities available to PhD holders, reflecting the broad applicability of their skills and expertise. The distribution highlights the evolving landscape of postdoctoral employment, emphasizing the importance of considering various career paths.

As life sciences continue to expand and diversify, so do opportunities for career development beyond the traditional academic path. In recent years, there has been a growing demand for trained professionals with expertise in the life sciences to fill various roles in industry, government, and other sectors.

This book aims to provide a comprehensive overview of the many career opportunities available to life science researchers outside academia. It offers guidance on leveraging one's scientific training and experience to best succeed in these roles. Additionally, the preparation of a resume (CV), cover letter, social media "attire" (LinkedIn), and insights on the hiring process will be covered.

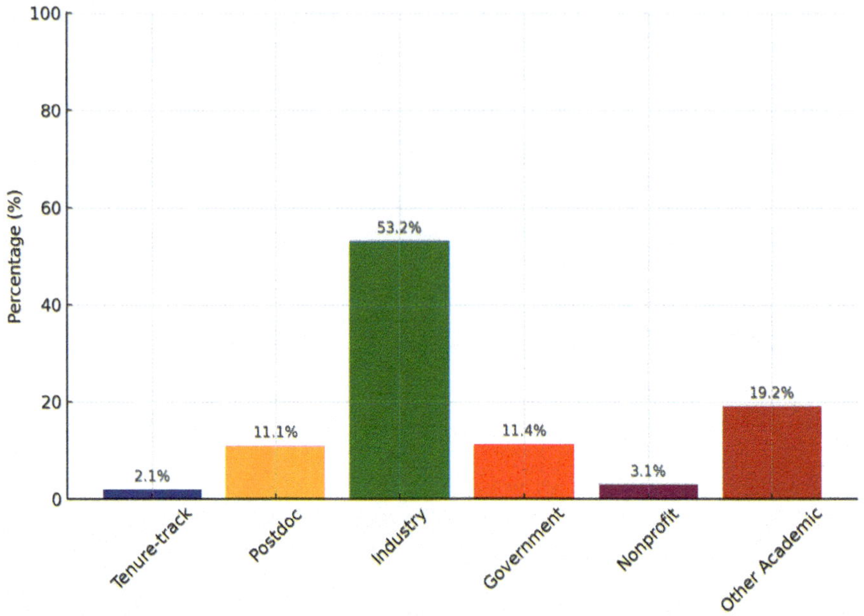

Fig. 1 Employment distribution of new PhD and Postdoctoral recipients by sector. This bar graph presents the percentage distribution of new PhD and postdoctoral recipients across various employment sectors. The data combines insights from the 2018 Taulbee Survey and the article "Visualizing detailed postdoctoral employment trends using a new career outcome taxonomy" to offer a comprehensive view of career paths in and outside academia. **Tenure-track (Blue):** This group represents individuals who have secured tenure-track faculty positions in academic institutions, highlighting a traditional and highly sought-after career path in academia. **Postdoc (Orange):** Indicates postdoctoral positions, a common next step for PhD graduates seeking further research experience, often in academic settings. **Industry (Green):** This category covers employment in for-profit companies, reflecting the appeal of the private sector for applying research skills in a commercial context. **Government (Red):** Includes positions within government agencies, showcasing the role of PhD and postdoctoral scholars in public sector research and policymaking. **Non-profit (Purple):** This category represents employment in non-profit organizations, where individuals contribute to research, advocacy, and humanitarian efforts. **Other Academic (Gray):** Accounts for academic roles that are not tenure-track, such as professional staff, management, or support roles within educational institutions, highlighting the diversity of career options within academia

The book also includes 20 insightful testimonials from researchers who started in Academia and transitioned to various roles in the Life Science Industry. The last part of the book will guide Universities or academic entities on how such an education model could be embedded in the standard curriculum to provide a full spectrum of opportunities to future people working in life science.

The authors have direct experience with the roles described in this book. They either worked in those roles themselves, hired, and managed for them or closely collaborated with clients and partners in those roles.

References

1. Edwards KA, Acheson-Field H, Rennane S, Zaber MA (2023) Mapping scientists' career trajectories in the survey of doctorate recipients using three statistical methods. Sci Rep 13:8119. https://doi.org/10.1038/s41598-023-34809-1
2. Xu H, Gilliam RST, Peddada SD, Buchold GM, Collins TRL (2018) Visualizing detailed postdoctoral employment trends using a new career outcome taxonomy. Nat Biotechnol 36:197–202. https://doi.org/10.1038/nbt.4059

Contents

About the Authors

Stefano Gaburro PhD, specializes in neuropharmacology and digital technologies for laboratory animal care. As Tecniplast's Scientific Director, he advances translational models for human diseases, emphasizing ethical research practices and the 3Rs—replacement, reduction, and refinement. A skilled communicator and Cialdini Certified Professional, Stefano leads significant webinars and collaborates globally to enhance animal welfare in scientific settings. His contributions are pivotal in shaping future biomedical research standards.

Eric Rieux boasts three decades of experience in high-value science-driven sales and marketing. Starting in 1992, he was instrumental in the growth of a US biomedical research instrumentation startup, eventually becoming VP of Sales & Marketing. From 2019 to 2022, Eric streamlined operations at a French life science software company. He recently trained in sustainable governance to help companies align with environmental and societal goals. Eric teaches at ESDES and INSEEC, enjoys the Alps, and is active in environmental advocacy.

1

Contemplating Professional Futures: Pondering Career Aspirations

Abstract This chapter encourages readers to reflect on their values, interests, and skills to identify fulfilling career paths. It emphasizes that aligning one's profession with personal aspirations is crucial for long-term job satisfaction and success. The chapter also guides utilizing personality profiling tools and understanding the transferable skills developed through scientific training, such as curiosity, organization, collaboration, creativity, observation, objectivity, and persistence. It concludes by emphasizing the importance of joining organizations that align with one's values and goals.

All we have to decide is what to do with the time given us.

J.R.R. Tolkien

We feel we have one unique, beautiful opportunity to lead a fulfilled life for ourselves and with all we are connected to: our families, friends, colleagues, communities, and environment.

The Greek philosopher Heraclitus is credited for saying, "Change is the only constant in life." However, based on our countless interactions with people of all cultural and educational backgrounds, we know one thing is constant: our inner personality. Who we are will evolve, but our true "self" will remain.

© The Author(s), under exclusive license to Springer Nature Switzerland AG 2024

S. Gaburro, E. Rieux, *Career Options in the Life Sciences: Guide Research for Your Path Beyond Academia*, https://doi.org/10.1007/978-3-031-73565-3_1

As this relates to your professional career and aspirations, ask yourself simple questions:

If I had a magic wand, what would the world look like? Which activities make me most happy? Do I like to work on a project independently or in a group? Do I enjoy going to new places and meeting people I don't know, or do I prefer to stay in one place and primarily work independently? Do I need to work on material items I can touch and feel, or do I like to work on more abstract tasks? What motivates me most? Helping others, inventing, creating, being in a leading or executing role? Do I enjoy change, am I comfortable with some unknown, or prefer stability? Do I strive best on small- or long-term projects?

We recommend you watch and reflect on Dr. Randy Pausch's last lecture at Carnegie Mellon University, Achieving Your Childhood Dreams [1].

We all face pressure from expectations from family, friends, spouse or partner, schooling, society, and even ourselves that "we should do what we went to school for." You got a Master's in Immunology, but working in a flower shop or being a plumber would be your dream and make you happy! Go for it! You will either regret it or do it anyway later in life!

There are several tools available for profiling personality; some are available free of charge, and others are available for a fee:

- Free: Enneagram Personality Test [2]
- Free: Based on Carl Jung's work [3]
- Free: Based on Myers-Briggs [4]
- Fee-based Myers-Briggs Type Indicator [5]
- Fee-based Insights [6]

The Indeed recruiting site has good career recommendations for given Myers-Briggs types [7].

Any test should be taken with a grain of salt to confirm what you already know, feel, and observe about yourself. We advise sharing and discussing test results with people who know you well.

You want to be aware of your strengths and preferences. It is no use applying for jobs that involve approaching people you don't know if this is very difficult for you and something you don't enjoy. Even if the company were to hire you, you would not be happy, and it would not last.

While every person has their qualities and personality, we have observed a standard set of assets that people develop as they study or work in science.

- Curiosity. Science is about knowledge and understanding mechanisms, having a theory or hypothesis, and proving it right or wrong. Why is that solution blue today? Why is that cell acting that way? Could I use fiber optics properties for something other than carrying data?

- Being curious brings innovation and challenges the status quo, improving companies. It also means learning continuously, an essential quality for adapting and growing in a constantly changing world.
- Organization. Designing and running experiments requires planning for supplies, timing, staffing, and funding.
- Being organized means that you will help the company plan ahead, stay on budget, you can work independently within a team, and can help others be organized.
- Collaboration. As a scientist, you likely have seen the power of collaborative work, either within a lab between technicians, students, researchers, or professors or between labs at the national or international level. Only with collaboration can we achieve together what we cannot do alone.
- Collaboration makes work efficient, rewarding, and fun. Great companies emphasize open communication and collaborative work.
- Creativity. Scientists mix using previous knowledge with trying new things. The only unlimited resource we have on Earth is our creative imagination.
- Creativity fuels innovation in products, services, and processes. Innovation is a critical component of company longevity.
- Observation and Objectivity. Science is based on facts, reproducible observations, and documented and peer-reviewed results. When a result is unexpected, it is noted as such. Action is taken to revise hypotheses and plan a new experiment.
- Observing and drawing an objective conclusion is precious in companies' fast paced, daily environment. This drives solid and logical decision-making.
- Persistence. Studies last weeks, months, years, or even decades. Staying the course takes remarkable courage, energy, and persistence.
- Being persistent and consistent builds resilience for companies. This is especially true for companies whose business models rely on continuously bringing innovative products to market or have long project cycles with their clients and partners (e.g., complex machines, processes, or software).

You may have heard or read that "any strength overused becomes weakness," which is true. If any of the above traits are taken too far, they could impede getting "stuff done":

Curiosity and Creativity should not lead to you questioning "everything everywhere every time." Most likely, there is a reason why things are done in a certain way or why a given process has seven steps. Questioning is good and healthy within reason.

Organization taken too far could impede agility. In industrial settings, the pace is often much faster than in Academia. There might be too much time to prepare, plan, and execute; this is "real life." Accept that life is not perfect and that success often comes from some form of negotiation and compromise. As the American philosopher and psychologist William James said, "The art of being wise is the art of knowing what to overlook."

Inclusive collaboration should not come at the price of delayed decision-making. Once you have gathered everyone's feedback, decide (if this is within your attribute) and move forward.

Being persistent does not mean being stubborn. This is when you must use your observational and people skills to judge the fine line between determined and obstinate.

Finally, it would be best if you aimed to join organizations that align with your values, interests, and personal goals. Chap. 5, "The Hiring Process," provides more information about this (Fig. 1.1).

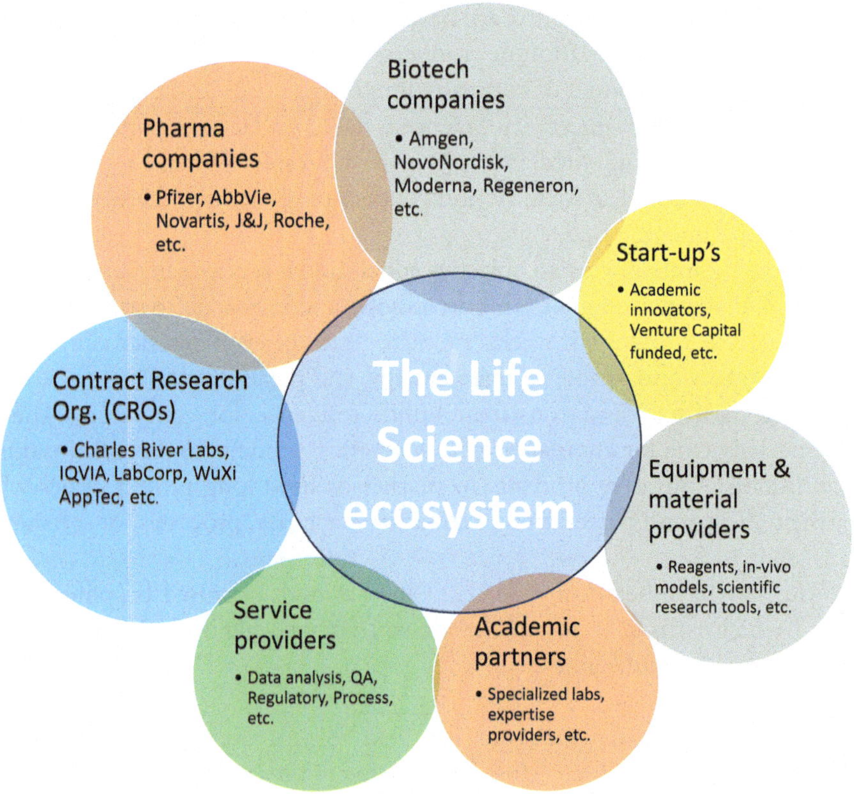

Fig. 1.1 Key stakeholders of the life science industry

Here is a partial view of the life science "ecosystem" with some of the key stakeholders:

- Pharma companies range from "Big pharma," with billions in annual revenue operating globally from R&D to commercialization, to smaller regional companies that may specialize in a given therapeutic area.
- Biotech companies focus on "biologics" (also known as "large molecules") to bring targeted treatments. They are often owned by or have merged with pharma companies.
- Contract Research Organizations ("CROs") are large global companies providing various services, from drug discovery to non-regulatory and regulatory pre-clinical and clinical testing. They serve Pharmaceutical and Biotech companies.
- Equipment and material providers and Service Providers range from smaller regional to global companies. They bring value to Pharma, Biotech, and CROs by providing highly specialized equipment and services.
- Academic partners are institutions and specialized labs collaborating with Pharmaceutical, Biotech, CRO, and possibly equipment and service providers… They provide unique or rare expertise in a given field.
- Startup companies are often started in Academia and based on a given discovery with a therapeutic application. The typical business model is to attract Venture Capital (VC) to prove the therapy's efficacy and then be acquired by a pharmaceutical or biotech company.

This is a limited list of the players and stakeholders within life science; one should research how and where each fits within the bigger picture.

References

1. Pausch R (2007) Randy Pausch last lecture: achieving your childhood dreams. https://www.youtube.com/watch?v=ji5_MqicxSo
2. Enneagram Personality Test (2018–2024) https://enneagram.bz/en
3. Idrlabs (2024) https://www.idrlabs.com/
4. 16 personalities (2011–2024) NERIS Analytics Limited. https://www.16personalities.com/
5. Myers & Briggs Foundation (2024) https://www.myersbriggs.org/
6. Insights (2024) The Insights Group Limited. https://www.insights.com/
7. Indeed.com (2024) Indeed Editorial Team, Updated October 13, 2023. The 16 Myers-Briggs personality types in the workplace. https://www.indeed.com/career-advice/career-development/myers-briggs-personality-types

2

The Value of Science-Educated Staff in the Life Sciences Industry

Our Candid Opinion About the Challenges and Opportunities of Academic Careers

Abstract This chapter delves into the challenges and opportunities of academic careers, focusing on academia's competitiveness and the crucial role of publishing and securing funding. It also discusses the valuable skills that academics bring to the life sciences industry, such as data analysis, research, technical writing, project management, and collaboration. The chapter emphasizes the industry's high regard for these skills and their ability to create numerous career opportunities beyond academia.

Academic life is characterized by high competitiveness, a scarcity of job opportunities, and significant pressure to publish research findings and secure financial support.

After completing a doctoral degree, individuals are fortunate to endure approximately 3–4 years if circumstances unfold favorably. Typically, transitions between different research environments are made to enhance one's reputation and acquire novel knowledge and techniques that can be disseminated and implemented across various laboratory settings.

Upon completing the initial year as a junior postdoctoral researcher, obligations related to teaching emerge, which are often imperative for advancing in academic circles but detract from research activities. Progressing through different professional positions leads to an increase in bureaucratic responsibilities and a reduction in laboratory hours.

Discussions with numerous colleagues have highlighted that the initial enthusiasm for making a global impact diminishes once individuals face bureaucratic challenges and the demanding cycle of publishing or perishing.

© The Author(s), under exclusive license to Springer Nature Switzerland AG 2024

S. Gaburro, E. Rieux, *Career Options in the Life Sciences: Guide Research for Your Path Beyond Academia*, https://doi.org/10.1007/978-3-031-73565-3_2

Securing tenure, commonly regarded as the ultimate achievement in an academic trajectory, demands prolonged and strenuous efforts. It necessitates substantial productivity and a proven influence on the field. Moreover, academic life frequently entails managing multiple roles, encompassing teaching, research, and contributions to the academic community.

Conversely, an academic vocation presents numerous prospects for personal and professional development, such as engaging in innovative research endeavors and collaborating with scholars globally. For individuals deeply passionate about their expertise, the opportunity to significantly contribute to advancing knowledge can be exceedingly fulfilling.

Furthermore, academic careers provide considerable flexibility and autonomy, enabling scholars to pursue their research interests and shape their professional journeys in alignment with their principles and objectives.

A career within academia equips individuals with diverse portable skills highly esteemed in the life sciences sector. Notably, individuals who have pursued academic paths have frequently cultivated robust competencies in critical thinking, problem solving, and data interpretation. They have also refined their capacity to work autonomously, oversee projects and resources, and convey intricate concepts to diverse audiences. These proficiencies are immensely transferable to the life sciences industry, where professionals frequently engage in research endeavors, data analysis, and dissemination of findings to various stakeholders. Moreover, academics are adept at collaborating with peers from diverse fields and backgrounds, a competence highly valued in an environment increasingly characterized by interdisciplinary cooperation. Overall, individuals who have pursued academic careers in the life sciences possess a distinct skill set that renders them highly sought-after candidates in the industry.

Below are some skills that can be transferred from academia to non-academic settings.

Data Analysis

Data analysis is a crucial proficiency in life sciences, enabling professionals to make well-informed decisions grounded on extensive data. A recent survey by the Society for Human Resource Management (SHRM) highlighted data analysis as the most sought-after skill in the life sciences sector for 2020. Professionals and researchers within the life sciences field are tasked with efficiently handling, structuring, and examining substantial amounts of data to extract valuable insights and draw meaningful conclusions. Individuals with a background in academia commonly possess significant expertise in data analysis, given that research endeavors entail collecting and analyzing extensive data sets. Their adeptness lies in utilizing statistical software tools and methodologies, data visualization, and data interpretation.

The application of data analysis in the life sciences domain encompasses various contexts, including drug development, clinical trials, and epidemiological studies. It empowers professionals to make decisions based on empirical evidence, discern patterns and trends, and uncover novel revelations. For instance, data analysis plays a pivotal role in drug development by scrutinizing the effectiveness and safety of a new pharmaceutical product, pinpointing potential side effects, and determining the optimal dosage. In clinical trials, data analysis assesses the efficacy of a novel therapy, identifies potential adverse events, and ascertains the most suitable treatment regimen for patients.

Data analysis proficiency necessitates solid quantitative skills, handling intricate data sets, and a profound comprehension of statistical principles and methodologies. Professionals and researchers with a background in academia are typically well equipped for such analytical undertakings, having engaged with complex data sets, conducted statistical analyses, and utilized statistical software tools. Their expertise often extends to data visualization and interpretation, indispensable skills for practical data analysis.

In summary, data analysis is a critical skill within the life sciences industry, and individuals possessing academic experience are aptly prepared for such analytical tasks. They have honed their abilities in managing, structuring, and analyzing extensive data sets and utilizing statistical software tools and methodologies. These competencies are pivotal in facilitating evidence-based decision-making, fostering product innovation, and propelling life science advancements.

Innovation and Curiosity

Research and development (R&D) is a crucial skill in the life sciences industry as it drives discoveries, treatments, and technologies. Those with experience in academia often have extensive experience conducting research projects, from hypothesis development to data analysis and dissemination of findings. They are proficient in designing experiments, using scientific equipment, and collaborating with peers.

In the life sciences industry, R&D professionals strive to develop new drugs, treatments, and medical devices and improve existing products. To do this, they must analyze project feasibility, design experiments, and conduct clinical trials. Effective R&D requires project management skills, teamwork, regulatory requirements, and knowledge of standards.

Individuals with academic experience are well-suited for R&D work in the life sciences industry. They have honed skills in project management, experimental design, and collaboration. They excel in analyzing data, using scientific equipment and instruments, and communicating findings to stakeholders. They are familiar with research's ethical and regulatory requirements, which is crucial in the life sciences industry.

Furthermore, academia prepares individuals with critical thinking, problem-solving, and creativity skills essential for successful R&D in the life sciences industry. They can think creatively and devise new ideas for product development or improvement. They are also skilled in identifying potential challenges and solving complex problems.

R&D is vital in the life sciences industry, and individuals with academic experience are well-prepared for this work. They are skilled in project management, experimental design, collaboration, critical thinking, problem solving, and creativity. These abilities are fundamental in developing new products, improving existing products, and driving innovation in the life sciences industry.

Technical Writing

It is a crucial competency within life sciences, enabling professionals to convey intricate scientific concepts to diverse audiences. Those with a background in academia often excel in technical writing, having authored research papers, grant proposals, and scientific reports characterized by clarity, brevity, and precision. Such individuals have cultivated expertise in scientific writing, data analysis, and disseminating findings to stakeholders.

Technical writers in the life sciences sector are tasked with producing regulatory documents like drug development reports and product manuals and disseminating findings to various recipients, including patients, clinicians, and scientists. Precision and brevity are paramount in this domain. Proficient technical writers must also possess adept editing capabilities and familiarity with style guides and formatting standards.

Academia-experienced individuals are aptly equipped for technical writing within the life sciences industry, having refined their proficiency in scientific writing, data interpretation, and impactful communication. They excel in structuring intricate data into lucid and succinct documents while simplifying technical content for diverse audiences.

Technical writers in the life sciences sphere must also demonstrate the ability to collaborate effectively with subject matter experts, regulatory affairs professionals, and other stakeholders to ensure documents align with regulatory prerequisites and are communicated efficiently. Collaboration and adaptability to evolving demands are pivotal for succeeding in technical writing within the life sciences sector.

In conclusion, technical writing is a pivotal skill in the life sciences industry, and those with academic experience are often primed for such roles. Their expertise in scientific writing, data analysis, and impactful communication allows them to structure complex information into clear and concise documents adeptly. Their proficiency in collaborative work with stakeholders to ensure regulatory compliance and effective communication is foundational to technical writers' triumph in the life sciences sector.

Project Management

Project management is vital in the life sciences sector, allowing professionals to oversee resources and finances to accomplish specific objectives efficiently. Those with a background in academia often have experience handling research projects, which involve coordinating team members, managing budgets, and adhering to timelines. They possess expertise in project planning, resource distribution, and objective establishment.

Within the life sciences field, interdisciplinary teams frequently consist of individuals with diverse skill sets, including scientists, clinicians, engineers, regulatory specialists, and business professionals. Effectively collaborating within these teams necessitates grasping varied viewpoints, proficient communication abilities, and adeptness in negotiation and conflict resolution.

Individuals with an academic background are well-matched for project management roles in the life sciences industry, given their proficiency in project planning, resource allocation, and objective setting. They excel in collaborating within interdisciplinary teams, negotiating and resolving conflicts, and adjusting to evolving demands. They also know research's ethical and regulatory standards, which are crucial in the life sciences sector.

Competent project management also demands robust leadership qualities, such as motivating and inspiring team members, delegating responsibilities, and providing feedback. Project managers must also excel in risk evaluation and management, recognizing potential challenges and formulating backup plans to tackle them.

In conclusion, project management is a critical skill within the life sciences industry, and individuals with a background in academia are frequently well-equipped for such roles. They possess expertise in project planning, resource allocation, objective setting, negotiation, conflict resolution, and adaptation to evolving requirements. They adeptly lead teams, identify and manage risks, and collaborate effectively with stakeholders. These proficiencies are fundamental to the success of project managers in the life sciences sector.

Collaboration and Teamwork

Collaboration and teamwork are fundamental competencies within academia and the life sciences sector. Individuals in these domains frequently engage with peers from diverse backgrounds and fields to undertake research, innovate products, and address intricate challenges. Within academia, researchers often collaborate with counterparts from various departments, universities, or even different nations to exchange knowledge, resources, and insights. Moreover, they may cooperate with industry affiliates, community entities, or governmental bodies to devise resolutions for practical issues. This exposure to collaborative endeavors can be readily translated to the life sciences industry, where professionals collaborate in interdisciplinary teams to introduce novel products, execute clinical trials, or scrutinize data.

Efficient collaboration and teamwork necessitate adept communication capabilities, mutual trust, respect among team members, and a readiness to share expertise and resources. Studies indicate that effective teamwork can enhance outcomes in the life sciences field, leading to superior product development, enhanced research quality, and heightened innovation levels (Dwyer, 2014). Collaborative teams are also more inclined to devise innovative answers to intricate problems and arrive at well-informed decisions.

Cross-functional teams within the life sciences sector may encompass individuals with diverse areas of proficiency, including scientists, clinicians, engineers, regulatory specialists, and business experts. Effectively functioning within such teams mandates an appreciation for varied viewpoints, strong communication proficiencies, and the capacity to negotiate and resolve conflicts. Researchers and professionals with academic backgrounds are typically well-equipped for such collaborations, having interacted with peers from different academic disciplines, backgrounds, and societal contexts. They often possess adept communication, negotiation, and conflict-resolution skills, crucial for successful collaboration and teamwork within the life sciences industry.

Collaboration and teamwork are indispensable proficiencies within academia and the life sciences sector. Given their honed communication, collaboration, and conflict-resolution skills, individuals with academic exposure are generally well-suited to meet the challenges of the life sciences industry. These competencies play a vital role in fostering fruitful collaboration and teamwork within the life sciences industry, where professionals unite their efforts to pioneer new products, conduct research, and address intricate challenges.

3

Current Roles in Life Science Industries

Abstract This chapter provides an overview of various roles available to life science professionals in the industry, including scientists in pharmaceutical or contract research organizations (CROs), scientific sales, marketing positions, application scientists, technical support, webinar or content creators, and scientific and medical writers. It describes the responsibilities, required skills, and potential career paths for each role, offering insights into the diverse opportunities available beyond traditional research positions.

Scientist Within a Pharmaceutical or Contract Research Organization (CRO)

This is one of the most likely opportunities someone with a life science degree can pursue. Most science-related skills are directly transferable: develop a research plan, perform bibliography reviews, learn and implement innovative methods, etc.

One of the key differences will be related to the purpose of the research. In industry, research programs are supported by private funds from investors and companies' profits. Research programs are regularly vetted and evaluated to ensure they are on track and will eventually pay off and bring a "return on investment." Speaking from experience, we have often seen quick decisions being made when significant liabilities were discovered for a promising drug or therapy. Bringing a new compound to market can cost over 2 billion dollars or Euros and take 10–12 years. The websites of the European Federation

© The Author(s), under exclusive license to Springer Nature Switzerland AG 2024

S. Gaburro, E. Rieux, *Career Options in the Life Sciences: Guide Research for Your Path Beyond Academia*, https://doi.org/10.1007/978-3-031-73565-3_3

of Pharmaceutical Industries and Associations (https://www.efpia.eu/) and the Pharmaceutical Research and Manufacturers of America (https://phrma.org/) provide many resources to understand the process of developing a new therapy. This is an investment-heavy and risky endeavor, so companies must diligently make strategic decisions.Other key differences relate to working environments. In academia, one usually has more leeway to work on one's agenda and wear "relaxed" physical attire. At the same time, in the industry, the pace is generally faster and the environment more formal (with significant variations from company to company). In academia, you may find yourself working with older equipment and limited resources, while in industry, you will have access to the latest and greatest and operate under an established budget.

One of the most significant differences is career and professional development opportunities. We have seen many of our scientist friends evolve into people or project management roles, transversal roles (quality, regulatory, clinical, etc.), or have opportunities to move to other locations nationally or internationally within the same company.

We want to make a special mention of "Contract Research Organizations" (CROs). Over the past 20 years, CROs have taken an ever-growing role in the Life Sciences ecosystem. Pharmaceutical companies, from start-ups to established "big pharma," rely on CROs to perform specific parts of their R&D process. As the process has become more complex and expensive, using specialized contractors has become more financially efficient, displacing internal costs to external and project-based expenses. For example, some CROs specialized in producing genetically modified mice models to study and perform early discovery studies. Others specialized in running regulatory Safety Assessment studies (Safety Pharmacology and Toxicology) before clinical trials. Others are yet dedicated to running clinical studies or can offer an "end-to-end" panel of services. CROs have been hiring many scientists to build up their capacity and can provide excellent job and career opportunities. Because the CRO arena is highly competitive and work is 100% project-based, you should expect a fast-paced working environment where you will likely face and feel pressure from sponsors before ("When can you start my study?"), during ("How is my study going? Give me preliminary data?") and after ("I need my study report from yesterday!") (Table 3.1).

Table 3.1 Hard and soft skills of scientists within a Pharmaceutical or Contract Research Organization (CRO)

Hard skills	Soft skills
Possess a scientific background	**Resilience, persistence, empathy**
Project management abilities	**Be a good listener**
Writing concise and efficient reports, proposals, messages	**Ability to work with multiple teams inside and outside your organization**

A typical day in the life of a pharma or CRO scientist:

- Plan or update a study timeline, ensuring staff and resources are correctly allocated.
- Provide updates on ongoing projects.
- Handle unplanned issues or deviations from the project plan.
- Talk to the different stakeholders in the project.
- Depending on seniority, perform experiments.
- Data analysis and reporting.

LinkedIn keywords examples of jobs: scientist, senior scientist, project manager, toxicologist, investigator

Scientific Sales

Do you have an outgoing personality, enjoy traveling, and can "sell"?

If this sounds like something you would be interested in, scientific sales is a non-academic profession in the life sciences business open to scientists. In this capacity, experts are responsible for promoting and selling scientific products and services to end users, including medical or research institutes.

Those who wish to be successful in scientific sales must have a comprehensive knowledge of the products and services they are selling and the requirements and preferences of their consumers. They must also explain complicated scientific ideas clearly and succinctly so potential clients can understand them.

Because successful scientific sales professionals are required to establish and maintain relationships with potential customers, identify their needs and preferences, and develop solutions that meet their requirements, successful scientific sales professionals frequently possess solid relationship-building skills. To effectively address the problems their clientele faces, they must be able to convey the benefits of their goods and services and comprehend the requirements of those clients.

Scientific sales professionals are expected to be knowledgeable about the most recent developments and trends in their field and the regulatory regulations and norms in place. They need to be able to provide technical help when needed and answer clients' technical questions.

Working in scientific sales can be a satisfying career choice for those with a scientific background. This line of work provides an opportunity to work with innovative products and technology while also contributing to the advancement of scientific research. To succeed in this profession, you must have strong communication and relationship-building abilities and an in-depth understanding of scientific ideas and methods. In our experience, relationships need to be built at multiple levels, from the technical staff who will be the daily users of the product or service, their managers who are concerned about efficiencies and job satisfaction, and other departments that may need to be involved such as Information Technology (IT), Quality Assurance (QA), and Finance for reporting, and of course, the people who will approve the purchase and implementation of what you promote (Table 3.2).

There is often a stereotypical view that a salesperson is someone who can talk well, impress people, and "sell anything to anyone." While this may be the case for low-value items sold to the public, scientific sales professionals are addressing educated, highly informed clients and are looking to develop long-term business relationships. Depending on the product, the sales cycle might vary from a few weeks for lower-value items to several years for complex systems involving many processes. So, it would be best if you were a self-starter, able to self-motivate, and often work "independently within a team." You will also face competition from other companies, and losing a project or a bid will

Table 3.2 Hard and soft skills of scientific sales professionals

Hard skills	Soft skills
Possess a technical or scientific background	Resilience, persistence, empathy
Writing concise and efficient reports, proposals, messages	Be a good listener
Ability to work with Customer Relationship Management (CRM) and planning software tools	Ability to connect to people and establish a network
Able to travel extensively to customer sites and conferences (may include weekends)	Be at ease with approaching and meeting people you don't know
	Excellent oral communication, possibly in 2 or more languages
	Ability to work with multiple teams inside and outside your organization
	Ability to think strategically for both your company and your clients

happen. Resilience, determination, and a positive outlook are essential to succeed in this role. If you have not guessed by now, both authors, Stefano and Eric, have been in this role and will comment that "these were some of the most fun and formative years of our professional lives!"

A typical day in the life of a scientific salesperson:

- Get up at 6 am to catch a train or plane or drive to a client in time for a meeting.
- Make 12 phone or video calls in 1 day to discuss new potential opportunities or stay in touch with clients.
- Spend a few hours writing (part of) a technical and financial proposal.
- Coordinate actions with your marketing, technical support, engineering, and management colleagues to prepare a large project for a client.
- Prepare a presentation and deliver it in front of a group.
- Handle severe customer-facing problems caused by the products or services you sell.
- Participate in scientific exhibitions, including a "trade show," and stay engaged for long days talking to potential and existing clients.

LinkedIn keywords for job examples: "technical sales specialist," "scientific sales," "scientific account manager," and "scientific business (or market) development manager."

Marketing Position: Product Management/ Market Management

Are you intrigued by what makes some products or services more successful than others? You may want to consider a marketing role as a product manager. In this capacity, scientists are responsible for developing and marketing scientific products or services. Product managers are accountable for ensuring that their companies' wares cater to the requirements and inclinations of their target demographic while also generating a profit for their respective companies.

Product managers must understand the scientific principles and methodologies relevant to their products or services and have excellent communication and leadership abilities to succeed in this profession. To successfully create and deploy their products, they must work effectively in teams spanning several functional areas, including scientists, engineers, marketing experts, and other stakeholders.

Successful product managers are typically proficient in project management, market analysis, and product creation. They can also recognize and respond to their customers' demands and preferences. They need to be able to adjust to different market conditions and collaborate with their team to develop new product strategies that will drive growth and increase profitability.

Product managers could be accountable for various goods and services, such as scientific tools, laboratory supplies, or software solutions. They must understand their products' technical specifications and the relevant regulatory requirements and recommendations. They must also convey to clients the value of their products clearly and concisely and devise marketing tactics that effectively reach their intended demographic.

Product management is a demanding and fulfilling career path for scientists in the life sciences business. It allows scientists to exploit their scientific expertise in a new and exciting way while also driving innovation and growth in the company.

Please note that many other marketing positions are suitable for science-educated people. For example, instead of managing a product or service, marketers can manage a specific market segment based on the type of applications, client type (minor, significant, academic or private, etc.), geography, or other criteria matching the company's marketing strategy. Marketers can also specialize in managing a commercial channel, for example, to support direct sales or sales via distributors (Table 3.3).

Table 3.3 Hard and soft skills of scientific product managers

Hard skills	Soft skills
Possess a technical or scientific background	Ability to manage projects within resources: people, budget, time
Ability to write concise and efficient reports, proposals, and messages	Be a good listener
Ability to work with spreadsheet and planning software tools	Ability to connect to people and work with multiple teams: sales, engineering, quality, finance, production
Able to travel to customer sites and conferences (may include weekends)	Ability to think strategically

A typical day in the life of a scientific product manager:

- Perform market research and analysis to define product specifications, value, which clients type to target, how, and where.
- Meet with colleagues in the engineering department to plan, review, and update on new product development or product improvement; or meet with colleagues in Production to check capacity status, lead times, quality, etc.
- Participate in scientific exhibitions, including a "trade show," and stay engaged for long days talking to potential and existing clients.
- Establish client beta sites and possibly visit them before and during the testing phase.
- Work with colleagues within marketing to develop promotional materials and campaigns.

LinkedIn keywords examples of jobs within "Life Sciences": "product manager," "scientific (or "biomedical") market manager," "marketing coordinator," "digital marketing manager," "product launch specialist."

Application Scientist

This position also closely resembles your actual occupation. Companies wishing to commercialize their product can use this to their advantage by transforming the story into one or more new scientific applications, which they can then use to market the product and the entire picture.

Application scientists are non-academic specialists who operate in a laboratory or other technical environment to provide scientific research assistance to clients. In this capacity, scientists can use their scientific knowledge by offering clients technical support, troubleshooting assistance, and training on using scientific products and services.

Account managers typically bring in Application Scientists at the pre-post-sales stage of a sales cycle. This allows account managers to have more productive conversations about the product's scientific uses with customers.

To succeed in this profession, application scientists need a thorough understanding of scientific ideas and methods and excellent communication and customer service abilities. They must effectively communicate complex scientific concepts to customers and collaborate to build solutions that fit each customer's research requirements.

Table 3.4 Hard and soft skills of application scientists

Hard skills	Soft skills
Scientific background Technical background (e.g., knowing how a scientific system works and understanding principles) Understanding common jargon in science Ability to write white papers or website content Analysis skills to potentially provide a biological interpretation of the data	Communication: being able to interface both with customers and with internal figures, e.g., engineers to improve the product Problem-solving: Strong analytical and creative problem-solving skills to innovate and solve challenges arising during scientific application and development Adaptability: Being flexible and adaptable in a fast-paced environment where technologies and project priorities may rapidly change

To be successful as application scientists for scientists, one must also know about the most recent developments and trends in the business, as well as the requirements and criteria set forth by regulatory agencies. They have to be able to respond to technically oriented customer inquiries and provide technical help when needed.

Application scientists for scientists, in general, play an essential part in the whole process of bolstering scientific research, promoting scientific products and services, and driving innovation within the life sciences industry. To succeed in this profession, you must possess good technical and communication abilities and an awareness of scientific ideas and methods (Table 3.4).

A typical day in the life of an Application Scientist:

- Discuss the needs of an existing or potential new client.
- Assist the sales team in preparing proposals and bids.
- Travel to a client's site to provide advanced consultation on a project.
- Work closely with Product Management and Engineering on the specifications and features of new or improved product(s) versions.
- Assist Marketing in articulating the value of a product or service, which will drive pricing.
- Attend scientific conferences and webinars to stay updated in your field of expertise and to present your company's product scientifically.

LinkedIn keywords examples of jobs: "application(s) scientist," "Field Application Scientist," "Field Support Scientist," and "Technical Field Specialist."

Technical Support

During your PhD studies, have you ever been required to set up new equipment or technology, validate it, or design anything to meet your specific requirements? But you are not necessarily interested in experimenting on your own? Technical support positions can offer a viable solution.

The life sciences business allows scientists to work in non-academic roles, such as providing technical support. Professionals fill this job by offering technical support and assistance to clients using scientific goods and services. In this profession, scientists can use their scientific expertise by troubleshooting and supporting customers whose scientific equipment or software is experiencing technical difficulties.

To succeed in technical support, scientists need a thorough understanding of scientific ideas and methods and excellent communication and customer service skills. They must be able to provide technical information to consumers clearly and succinctly and assist customers in resolving any technical issues they may have promptly and effectively.

The technical support expert should also be knowledgeable about the most recent business developments and trends and the needs and criteria set forth by regulatory bodies. They need to be able to provide technical help when needed and answer technical questions from clients.

Those with a scientific background can leverage their skills and expertise to provide excellent technical support to customers in the life sciences business, a vital job in the industry. To succeed in this profession, you must have solid technical and communication abilities and an awareness of scientific principles and procedures.

When clients call Technical Support, it is because they need help, and they need it now. They may be in the middle of a study or process that has stopped and face immense pressure to deliver a project on time. While most people are courteous and understanding, some will have little patience or will not speak their native language. A high dose of patience and empathy is required. Some problems may not be solved immediately, requiring you to enlist others to help (Table 3.5).

Table 3.5 Hard and soft skills of technical support specialists

Hard skills	Soft skills
Technical or scientific background	**Be a good listener**
Ability to use CRM and ticket management software tools	**Show and communicate empathy**
	Be resilient
Ability to write concise reports	**Work with other departments such as sales, engineering, quality**
Ability to do some travel to client sites	

A typical day in the life of a Technical Support Specialist:

- Answer 12 phones or video calls of customers facing technical issues.
- Perform product training remotely.
- Travel to clients to provide on-site training or troubleshoot and fix technical issues.
- Work with colleagues in Sales to best assist clients and with Marketing and Engineering to collaborate on improving products and developing new ones.

LinkedIn keywords are examples of jobs: "technical support specialist." "Technical Service Representative," "Customer Support Scientist," and "Technical Support Scientist."

Webinar or Content Creator

Webinar creators are professionals outside academia specializing in creating and delivering educational or promotional webinars, particularly in science or the life sciences industry. These webinars serve various purposes, such as marketing products, educating audiences, or showcasing scientific offerings. Scientists can leverage their expertise to develop webinars presenting research findings, discussing advancements in their field, or advertising their scientific services.

To excel in this role, professionals creating webinars for scientists must possess strong skills in content development, presentation, and effective communication. They should create visually appealing and informative presentations that are easy to comprehend, engaging, and informative. Moreover, they must simplify complex scientific concepts for diverse audiences, ensuring clarity and brevity in their delivery.

Successful webinar creators for scientists must collaborate effectively with a team of professionals, including marketers, technicians, and designers. They need to adapt to changing requirements, manage time constraints, and oversee project scope to produce impactful webinars for scientists. These creators are crucial in disseminating scientific knowledge, advancing research, promoting scientific products, and fostering innovation in the life sciences industry. To thrive in this field, individuals must possess strong technical skills, practical communication abilities, and a solid understanding of scientific principles and processes (Table 3.6).

A typical day in the life of a Content Creator:

Table 3.6 Hard and soft skills of webinar or content creators

Hard skills	Soft skills
Scientific knowledge: Strong understanding of relevant scientific topics	Communication: Clearly explain complex topics and coordinate with various participants
Research skills: Ability to comprehend and summarize scientific research	Organizational skills: Manage scheduling, coordination, and technical setups efficiently
Content creation: Proficiency with software for creating text, images, and videos	Problem-solving: Quickly address and resolve any issues that arise during the webinar
Information gathering: Skills in compiling details like speaker coordination and webinar logistics	Interpersonal skills: Build strong relationships with speakers, attendees, and collaborators
Technical proficiency: Expertise in using webinar platforms and tools effectively	Adaptability: Adjust to new technologies, content changes, and audience needs
	Attention to detail: Ensure all webinar elements are polished and professional
	Patience and calmness: Stay composed and cheerful, even under stress

As a scientific webinar creator expert, your day would be varied and dynamic, focusing on the content and logistics of hosting webinars. Your morning might begin with a review of your goals and the day's agenda, ensuring that everything aligns with the key messages you intend to convey. This is followed by a period dedicated to refining and developing the webinar's content, including creating slides and interactive elements to engage the audience.

A technical check of all equipment and software would be essential to avoid hiccups during the live session. As the day progresses, you would meet with guest speakers or collaborators. These meetings are crucial for aligning everyone's contributions and conducting any necessary rehearsals.

Promotional activities also play a significant part in your day, involving updates on social media and other channels to ensure maximum participation. The afternoon might be spent finalizing the interactive components of the webinar, like polls and Q&A sessions, ensuring they are ready to go live.

The highlight of your day would be hosting the webinar—introducing speakers, moderating discussions, and ensuring the session runs smoothly. After the webinar, your responsibilities continue with post-event tasks such as responding to attendee questions, sending out follow-up materials, and collecting feedback to improve future webinars. This routine allows you to combine your scientific expertise with outreach and education, making every webinar a valuable learning experience for participants.

LinkedIn keywords examples of jobs: Webinar creator, online content creator

Scientific and Medical Writing

Scientists can apply their scientific knowledge and acquire proficiency in writing, communication, and critical analysis to transition into roles such as life science journalists, medical writers, or professionals in scientific journals. This transition can be achieved through the development of these essential skills. In these positions, individuals are tasked with composing articles or reports on scientific studies and progress and simplifying complex scientific concepts for the general public or fellow scientists in the same field.

Success in these professions for scientists hinges on a profound grasp of scientific principles and methodologies, coupled with adept writing and communication skills. They must be capable of conducting thorough analyses of scientific research and presenting it engaging and comprehensibly. Collaboration with other professionals, like editors or fellow writers, is crucial for delivering top-notch content.

Scientists can leverage transferable skills in these roles, including project management, attention to detail, and research capabilities. They must juggle multiple writing tasks simultaneously, ensuring timely and budget-friendly completion of each project. Furthermore, researching to gather information for their work requires a meticulous approach to guarantee accuracy and precision in their writing.

In essence, the roles of a life science journalist, medical writer, or scientific journal professional offer scientists a platform to apply their scientific expertise in a fresh and captivating manner, contributing to the dissemination of scientific discoveries. Those excelling in these fields possess a blend of scientific understanding and writing prowess, enabling them to create content that is both informative and engaging (Table 3.7).

A typical day in the life of a Scientific/Medical writer:

The day usually starts with a comprehensive review of the latest developments in medical research, including new studies, clinical trial results, and updates from health regulatory agencies. This professional stays abreast of advancements and changes in the field to ensure the accuracy and relevance of their work. Engaging with these materials, they meticulously sift through technical data and statistics to distill the essential elements needed for their writing projects.

Table 3.7 Hard and soft skills of scientific and medical writers

Hard skills	Soft skills
Scientific knowledge: Deep understanding of the relevant scientific and medical fields	**Attention to detail: Precision in capturing scientific data and findings**
Technical writing: Ability to convey complex information clearly and accurately	**Communication: Effective in explaining complex concepts to diverse audiences**
Research proficiency: Skilled in literature review and data interpretation	**Time management: Efficient in managing deadlines and project timelines**
Document formatting: Knowledge of industry standard formats and regulatory requirements	**Adaptability: Flexible in handling various writing tasks and audience needs**

The bulk of a scientific medical writer's day is devoted to crafting documents ranging from research papers and grant proposals to educational materials for healthcare providers and patients. Precision in language and factual accuracy are paramount, as the content must adhere to strict medical standards and be accessible to non-specialists without compromising the depth of information. Collaboration is a vital part of their routine; they often coordinate with researchers, healthcare professionals, and editorial teams to align on the narrative, resolve discrepancies in data interpretation, and refine drafts. They fulfill a crucial function in disseminating medical knowledge by bridging the gap between complex scientific data and practical application.

This chapter lists the most common positions within life sciences for PhDs and postdocs, but many others are "niche" yet rewarding and exciting. You can read some examples in Chap. 5, "Testimonials," here to name a few:

- Life Sciences PhDs can find career opportunities in publishing. This can mainly be for Scientific Journals and other magazines that regularly publish popular articles.
- Positions within NGOs or Foundations active within the research/medical world.
- Positions within Life Sciences companies' Regulatory, Quality Assurance, and Compliance departments.
- Position within production facilities of Pharmaceuticals or Biotech companies.

4

Tips and Strategies for Finding Job Opportunities: Resume and Online Presence on LinkedIn

Abstract This chapter offers practical advice on transitioning from academia to industry. It focuses on transforming academic CVs into impactful industry resumes and building a strong online presence on LinkedIn. It provides detailed guidance on crafting a compelling resume summary, highlighting transferable skills, and showcasing relevant experience. The chapter also emphasizes the importance of networking and utilizing LinkedIn to connect with potential employers and explore job opportunities.

Transforming Your Academic CV into a Powerful Industry Resume

For academics transitioning into life science industry roles, your extensive academic CV alone is insufficient for getting hired. Unlike faculty positions where comprehensively documenting your scholarly journey is expected, industry employers spend only 15 s initially screening resumes. In this rapid first pass, your CV's dense pages of publications, conference presentations, and academic accomplishments cannot capture their attention. You need a strategic, condensed marketing document laser-focused on the specific qualifications they require for that role.

As someone who successfully transitioned from over a decade in academia researching pharmacology, animal models, and behavioral neuroscience to leadership roles at major scientific companies like Tecniplast, Harvard Biosciences, and smaller pioneering firms, I know firsthand how difficult it

© The Author(s), under exclusive license to Springer Nature Switzerland AG 2024

S. Gaburro, E. Rieux, *Career Options in the Life Sciences: Guide Research for Your Path Beyond Academia*, https://doi.org/10.1007/978-3-031-73565-3_4

can be to repackage years of experience. When I began exploring new opportunities, I quickly realized my lengthy CV made it nearly impossible for employers to gauge how my background translated to their priorities rapidly. That is when I committed to transforming my CV into a powerful two-page industry resume using the following strategies:

Craft an Impactful Summary Statement Replace the traditional CV objective with 3–4 standout bullet points that encapsulate your most valuable technical skills, quantify years of experience, and highlight alignment with the company's particular focus. This section captures attention by demonstrating your authority and credibility as an expert fit for the role.

For example, "Research scientist with over eight years developing and characterizing in vivo models of neurological disorders like epilepsy and Alzheimer's disease using cutting-edge surgical techniques and state-of-the-art monitoring equipment." This creates a knowledge shortcut, signaling you as an experienced, qualified candidate.

Articulate Tangible Technical ExperienceUtilize active phrasing to highlight specific methodologies and impress with quantifiable outcomes like publications, funding, and process improvements. "Led electrophysiology studies in rodent models of anxiety, resulting in 4 co-authored publications in prestigious journals and a $300K grant award for pioneering new deep brain stimulation protocol."

By providing vivid, concrete details and social proof, such as funded grants and peer-reviewed work, you increase your perceived authority and trustworthiness as someone who can deliver results in this domain. This makes your background more memorable and impactful.

Isolate Technical Skills and Core Publications Create a distinct "Skills" section that comprehensively lists all the required expertise mentioned in the job posting using the same terminology. This will demonstrate your intricate knowledge of the specialized tools and concepts essential for success.

For example, "In vivo disease models (neuroinflammation, neurodegeneration, psychiatric), neurobehavioral testing, molecular/biochemical techniques, data analysis (SPSS, Prism, MATLAB), surgical techniques (stereotaxy, telemetry implantation)."

Seeing these tangible capabilities transparently listed activates the contrast principle—it provides a precise reference point for comparing your skills against the role's requirements. This contrast increases your perceived superiority as a subject matter expert candidate.

Finally, a "Selected Publications" section showcases your technical expertise through peer-reviewed work directly in the role's domain. You have provided multiple sources reinforcing your authority on the required specialized knowledge.

You leverage psychology principles like authority, social proof, contrast, and clarity by strategically repackaging your academic achievements into an ultra-tailored snapshot. Your background is framed most compellingly as distinctly qualified to provide immediate impact in that specific industry setting. The science of influence tells us that tapping into these persuasive tactics subtly shapes how resume reviewers perceive and evaluate candidates.

As respected scholar Robert Cialdini highlights in his book Influence, we are hardwired to follow shortcuts that convey credibility cues. By making your skills unmistakable through transparent skill lists mapped to the role, quantifying accomplishments through undeniable metrics, and providing sources corroborating your authority, like publications and credentials, you gain an edge that bypasses critical scrutiny. The hiring manager develops an instinctive positive feeling from this "professionalism heuristic" of recognizing an evident subject matter expert.

However, while these techniques are powerful, they should never misrepresent your true capabilities. The goal is compellingly reframing your rhetoric and structure to resonate optimally with screeners' decision-making tendencies. Once past the initial gating, you must still substantively prove your worth through technical interviews, portfolios, and vigorous conversations about applying your experience to their strategic objectives.

While transforming a comprehensive CV into a punchy, customized resume takes effort, it allows you to ethically harness principles of influence and compellingly position yourself as the ideal candidate. This unlocks opportunities to showcase your passion and potential as a valued contributor in life science industry roles.

Your Online Presence on Social Media: LinkedIn

Your transition from academia to industry is an exciting adventure with new possibilities. As you embark on this journey, your LinkedIn profile becomes a powerful storytelling tool, allowing you to share your unique experiences and expertise with potential employers. This section will guide you in crafting a compelling narrative that bridges your academic accomplishments with your aspirations for the future.

The Banner represents your statement, and it helps set the tone. The banner image is the first impression on your profile, setting the tone for your professional story. Consider a picture that resonates with your new career goals, whether it is a symbolic representation of your field, an inspiring landscape, or an abstract design that aligns with your values. This visual introduction

Fig. 4.1 Stefano Gaburro screenshot of a successful banner for LinkedIn profile. The banner contains all information needed for the visitors needed

should pique curiosity and invite others to explore your narrative further. You can write like in my example in three points what I do (Fig. 4.1).

Profile Picture: A Friendly Introduction Your profile picture is an opportunity to make a warm, personable connection. Choose a professional yet approachable headshot that reflects your genuine self. Dress in a manner that aligns with the professional standards of your target industry, making a positive first impression. When you comment, remember you are on mobile or a PC, a small ball. Try a headshot with a plain background (similar background color to your banner).

Headline: Showcasing Your Unique Value Your headline is more than just a title; it is a chance to encapsulate the value you bring to your new professional realm. Craft a concise yet compelling headline highlighting your expertise and the transferable skills you have honed through your academic journey. Depending on your audience, you might need to create authority (like in my case) or write about your skills (e.g., I help scientists moving out of academia via….).

About Section/Info: Weaving Your Professional Tale The About section is where you can genuinely narrate your transition from academia to industry. Share your motivations, aspirations, and the unique perspectives gained through your academic experiences. Emphasize the versatile skills you have developed, such as analytical thinking, project management, and effective communication. Weave in relevant keywords to improve visibility and demonstrate your alignment with industry terminology. This section could disappear shortly and become less relevant.

Experience Section: Highlighting Transferable Skills In the experience section, showcase how your academic roles and achievements have equipped you with invaluable skills that translate seamlessly into your new professional path. Highlight notable projects, collaborations, and responsibilities, emphasizing the transferable talents you have cultivated, such as problem-solving, leadership, and adaptability. Quantify, e.g., I supervised many PhDs and won many grants. Quantifiable objects can be easily transferred. Now, you can assign the skills you used to each position.

Skills and Endorsements: A Testament to Your Capabilities Thoughtfully select skills that resonate with your new career aspirations, encompassing hard and soft skills you have acquired throughout your academic journey. Encourage colleagues, mentors, and peers to endorse your abilities, adding credibility and depth to your profile.

Recommendations: Personal Endorsements Recommendations serve as testimonials of your character, expertise, and professionalism. Seek recommendations from diverse connections, including academic mentors, colleagues, and any industry professionals you have collaborated with. These endorsements provide a well-rounded perspective on your capabilities and work ethic.

Customized URL: Your Professional Branding A customized LinkedIn URL enhances the professionalism and memorability of your online presence. Choose a name variation or a professional moniker that makes your profile easily identifiable and accessible.

Education and Certifications: Your Intellectual Foundation Your academic credentials and certifications form the bedrock of your professional identity. Proudly showcase your degrees, relevant coursework, and industry-specific certifications or training, highlighting your commitment to continuous learning and professional growth.

Accomplishments and Projects: Celebrating Your Achievements In this section, you can showcase your noteworthy publications, projects, and honors, underscoring your contributions to your field. For those transitioning out of academia, emphasize collaborative projects with industry partners or practical applications of your research, demonstrating your ability to bridge theory and real-world impact.

Volunteer Experience: Demonstrating Character and Commitment Including volunteer roles and experiences illustrates your character, work ethic, and dedication to community service. Highlight leadership positions, project management responsibilities, or any roles that allowed you to develop skills relevant to your new professional path.

Engagement: Building Your Professional Network An active presence on LinkedIn—through sharing insightful content, engaging with posts, and participating in discussions—demonstrates your genuine interest and involvement in your professional community. Regular engagement can increase your visibility, establish your reputation, and foster valuable connections.

Consistency and Completeness: The Hallmark of Your Profile Ensure that every section of your LinkedIn profile is complete, up-to-date, and presents a consistent narrative. A comprehensive and cohesive profile enhances your professional appeal and visibility in search results, making you more discoverable to potential employers.

Posting and Connection: Start commenting on peers' posts, follow them, and write them a 300-character invitation, not blank but explaining why you would love to connect with them. Is that a topic or a paper? Expand your network.

Below is what you could post based on the 2024 LinkedIn algorithm change. At its core, the 2024 update prioritizes and encourages substantive engagements and the dissemination of knowledge. To leverage the advancements in this algorithm, it is advised to concentrate on producing valuable content that promotes authentic relationships and is tailored to the interests and capabilities of your target audience.

Always try to add a graph, image, or picture to accompany your post (not needed for commenting). This addition increases the post reach considerably. Figure 4.2 shows the optimal content structure for the three different post formats.

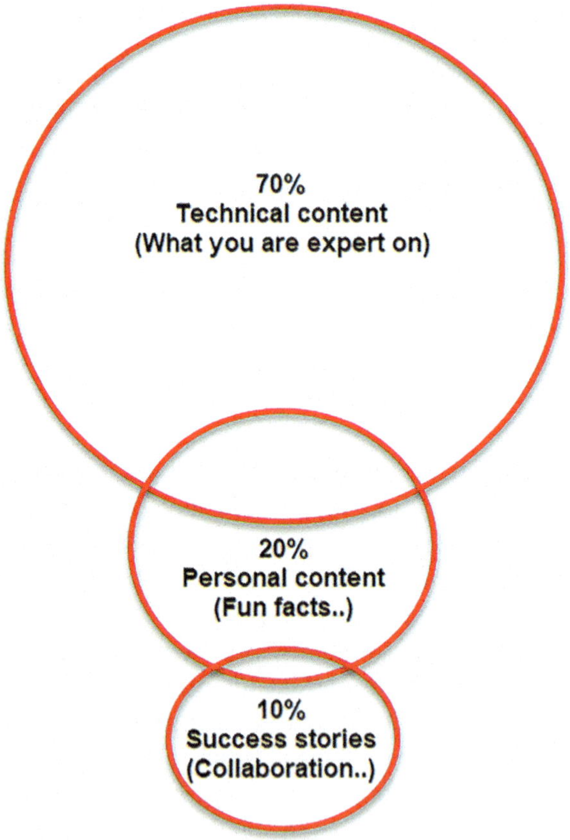

Fig. 4.2 How is the LinkedIn algorithm working? For successful posting the current algorithm ways 70% technical contacts, 20% personal content and 10% success stories (e.g., collaboration)

Your journey from academia to a new professional realm is a transformative adventure filled with opportunities for personal and professional growth. By thoughtfully crafting your LinkedIn profile, you can share your unique narrative, showcase your versatility, and demonstrate your readiness to embrace new challenges beyond academia. This intentional approach will position you as a dynamic, well-rounded professional poised for success in your chosen industry.

5

The Hiring Process: A Two-Way Street to Success!

Abstract This chapter explores the hiring process from both the employer's and the candidate's perspectives. It discusses the importance of aligning personal values with organizational values, understanding different corporate structures, and researching potential employers. The chapter also outlines the typical steps in a recruitment process, from resume screening to interviews and reference checks, offering tips on navigating each stage successfully. It emphasizes that the hiring process is a two-way street, where both parties assess compatibility and potential for a long-term working relationship.

There are countless resources and tips on "getting hired." However, engaging in a working relationship is not a one-way street where just the company "hires" employees. You, as a candidate, have the power to select your employer. A successful working relationship that will hopefully last several years considers both sides, the organization and the individual.

As we discussed in the first chapter of this book, finding a fulfilling role should ideally match one's personality, values, interests, skills, and competencies.

Our recommendation is to start with values. Most organizations communicate their values (core guiding principles), their mission (purpose, intention, objectives), and their vision (what will be achieved, what issue will be solved). This can be found on websites, commercial documents, and financial reports. Do you see these as inspirational and aspirational? If not, we know by experience that most working relationships will not last long if they are not aligned, at least in broad terms. We all work for money, but beyond that, a

© The Author(s), under exclusive license to Springer Nature Switzerland AG 2024

S. Gaburro, E. Rieux, *Career Options in the Life Sciences: Guide Research for Your Path Beyond Academia*, https://doi.org/10.1007/978-3-031-73565-3_5

common purpose is what keeps us engaged, energized, motivated, and, yes, happy! It may still be to your benefit to take a job you are not 100% excited about to gain a "first experience" or if you are willing to make some compromise for personal reasons. In this case, take the decision entirely, make the best of it, and remember that it is usually easier to find job opportunities while employed and connected.

Besides values, mission, and vision, consider the types of organization or corporate structure in which you are considering working. While this is not all black and white, here are some general considerations:

First, size matters. In small organizations, it is generally expected that you will have to do more than just what you were hired for and what is listed on your work contract or job description. This can be challenging and… rewarding. You may have more opportunities to try things, fail and try again, take on responsibilities, and provide ideas and insights compared to a larger company. Roles are generally more structured in larger companies (>300–500 employees). You will likely have more predictable and stable tasks and missions and possibly fewer opportunities to provide your ideas and views and make a visible contribution. On the other hand, you will likely have access to broader resources (people, budget). You may also have more opportunities to evolve into other roles in a different company department or geography.

Second, consider the organization's governance and financial structure: is the business owned by the initial founder(s), by a family, a private equity group, or listed on the stock market? Try to research and understand who the principal shareholders are. Are they aligned with the hiring organization's mission/vision/values and your values? Are they primarily interested in short-term financial gains or investing long term? Public companies are at the mercy of stock market swings and are under pressure to provide "positive" quarterly results. That can impact operating expenses, hiring, investments, etc.

Finally, research should be conducted on how potential employers treat their employees: Is the company growing? What is their employee turnover versus the industry average? How do they fare on equality questions? What kind of benefits do they provide? Use resources like LinkedIn, platforms like Glassdoor, and information you can get from your network.

The Recruitment Process

All too often in the past, and still valid for some organizations, people are recruited based chiefly on hard skills and competencies. While the technical aspects of a science-related job are essential, the interpersonal skills are just as or even more critical. Most of us can relate to a situation where someone cared for our needs but left us feeling poorly treated or considered. Likewise, one can be the best scientist in the world yet lack empathy, listening, presentation, etc., which are essential for many teamwork roles. We invite you to assess your personality traits (see Chap. 1) and make a matched comparison with the soft skills described in the various roles described in this book. If you are applying for a position and the potential employer is focusing on hard skills, you can bring up that topic and discuss the value your soft skills can bring.

Here, we want to share the typical steps you can expect from a recruitment process. More and more organizations now outline how they conduct their recruitment process. If not, here are the most common processes in use. Of course, there are many variations of what follows.

You will be asked to provide a CV (or resume) and possibly a cover letter explaining your interests and motivation. We refer you to Chap. 4, which offers tips and resources for CVs, LinkedIn profiles, etc. In our experience, the Human Resources department will do the first CV screening, looking at competencies, knowledge, and motivation in the cover letter. When more than 15 or 20 applications are received, be aware that only a few minutes are spent on each application, as low as 2 min even!

Then, HR will provide a shortlist to the manager who is making the hiring. At this point, the manager may refine the applicants' shortlist and involve some colleagues on the team to review applications.

The next likely step is to make the first contact via phone or video. Ideally, this will be a dialogue where the recruiter will want to get to know you and understand your motivation, interests, and competencies, during which you will ask questions. We recommend focusing that first call on the company's mission and the group you would be in, expectations for the role, questions about the team you would be in, etc. We expect that you have done some research before applying for the position, so now is the time to share possible ideas and contributions you could make. We recommend against asking questions like salary, benefits, and vacation time during this first call unless this is brought up.

This first call is about connecting at the personal level and establishing a first level of trust. Each side asks itself many questions: Can I trust this person? Do I want to talk to them again or meet in person to know more? Could I work with that person every day of the week? Will that person fit in my team? Do I feel good about being managed by this person?

Suppose you are still interested after this step. In that case, we recommend you send a short note to thank people for their time, summarize the exchange, and express your interest in pursuing the opportunity. It seems like an obvious step, but in our experience, less than 50% of candidates take the time to send a thank you letter! This is an easy way for you to stand out.

When the recruiting manager or HR manager completes all first phone interviews, the shortlist will likely be refined to a small number of candidates (2–5), and the process moves to a face-to-face interview. In our experience, face-to-face meetings are the best way to continue connecting on a personal level, strengthen the establishment of trust, and mutually explore what a future collaboration could look like. Observing the recruitment process in place will tell you a lot about the organization: are people well organized and on time? Did they communicate the logistics and expectations for the interview? Are they trying to put you at ease or under pressure? In our experience recruiting sales, technical support, marketing, and service positions, we used informal dialogue and role-play exercises. We also gave candidates a minor assignment ahead of time, leading to a presentation during the face-to-face interview. That allowed us to gauge motivation (did the candidate do a quick or thorough research?) and overall presentation skills (verbal and non-verbal communication). We also asked candidates to describe what they thought their first 6–12 months would be like. This was to evaluate whether the candidate understood our expectations (if not, we knew we had to reformulate) and if the candidate could project themselves into the role and mission. We often had two people from the company for interviews; this was meant to keep the dialogue flowing and to have one person observe non-verbal cues while the other was talking.

Another step many organizations take is to ask for and check for references. These can be from a peer, former manager, teacher/professor, or someone you have worked with in a non-profit setting. In our experience, we used references for two reasons: first, to confirm our impression of the candidate, and second, to obtain valuable insights to help future employees be at their best. For example, a former manager or colleague might share more about the candidate's preferred communication style.

Negotiating

People often ask us, "What can be negotiated?" The answer is simple: just about anything, as long as it fits within "reasonable boundaries." Salary is the main item, and benefits such as medical insurance, paid vacation, and sick time are often standard depending on the country you live and work in. "Reasonable boundaries" means asking for items or conditions within the range of what other employees of similar levels/roles get—for example, I can allot remote work and participate in events or conferences.

Keep in mind that a work relationship goes both ways. Hiring staff costs money, time, and energy, as do onboarding and training. In our experience, a new employee usually takes 6–12 months to be fully "productive" and "at ease" in their role. Organizations generally want to hire for the long term and essentially offer compensation and benefits in line with the local labor markets. They must also be fair to current employees and align with internal pay scales.

At the end of the day, you might not be able to negotiate much unless you have extraordinary qualities, skills, connections, etc. The best recommendation is to do your "homework" ahead of time so you have a good idea of the value you bring and are self-confident when it is time to discuss compensation and benefits.

Navigating New Horizons: The Role of Headhunters in Transitioning to Industry

Transitioning from academia to the life sciences industry can be daunting, filled with unknowns and challenges that require a shift in mindset and a strategic approach to career development. This is where the expertise of headhunters, or executive recruiters, becomes invaluable. Specializing in identifying and placing talent in specific industries, headhunters play a crucial role in bridging the gap between the unique skills of academic professionals and the particular needs of companies in the life sciences sector.

Headhunters are more than just recruiters; they are career advisors and industry insiders who understand both the explicit requirements of their client companies and the implicit nuances that contribute to a successful placement. Engaging with a headhunter can provide several advantages for academics looking to enter industry roles. These professionals offer access to unadvertised job opportunities and provide insights into the industry culture. They also offer resume and interview coaching and personalized advice on career development tailored to align an academic's expertise with market needs.

This subchapter will explore how academics can effectively collaborate with headhunters to navigate their entry into the life sciences industry. We will delve into the process of finding the right headhunter, what to expect during the recruitment process, and how to leverage their expertise to secure a position that best fits an academic's career aspirations and skills. Through practical advice and real-life examples, this section aims to equip readers with the knowledge and tools needed to make a successful transition with the aid of a professional recruiter.

We, therefore, based on our experience, report 10 critical points when dealing with headhunters:

- **Differences Between Academic CVs and Industry Resumes:** When academics apply for industry roles, they must adapt their CVs to be more concise and relevant. Academic CVs typically list all publications, academic presentations, and detailed research projects, which can be too exhaustive for industry applications. Instead, the focus should be shifted toward showcasing applicable leadership, project management, and problem-solving skills. **Example:** An academic supervising postgraduate students can emphasize their role in developing research talent and managing complex projects, which mirrors team leadership and project oversight in industry settings.
- Moreover, industry resumes should be results-oriented, highlighting outcomes and impacts rather than just activities. This involves quantifying achievements where possible, such as stating the percentage increase in funding due to a grant application or the efficiency improvements in a research process. **Example:** Instead of simply listing involvement in grant writing, an academic could specify that they successfully secured funding worth $500,000, leading to an expansion of research capabilities and a 30% increase in published papers.
- **Understanding Applicant Tracking Systems (ATS):** Applicant Tracking Systems (ATS) are used by many companies to manage the high volume of applications they receive. These systems scan resumes for keywords and phrases that match the job description, ranking each resume based on how well it aligns with the job requirements. Academics must understand the importance of including relevant keywords from the job posting in their resumes to ensure their application gets noticed. **Example:** If the job requires expertise in "cell culture techniques," it is crucial for the applicant to include this exact phrase in their resume.
- Additionally, formatting plays a significant role in ensuring the ATS can read a resume effectively. Academics should avoid headers and footers, use standard fonts, and keep the layout simple. Bullet points can be used to break down experiences and skills. **Example:** Instead of embedding important information in headers or using complex tables, listing experiences in a clear, bulleted format ensures that critical details are easily accessible not

just to ATS but also to human reviewers, making it more likely that the resume advances to the next stage of the hiring process.

- **Valued Skills in the Life Sciences Industry:** In the life sciences industry, specific skills that may not be emphasized in academia are highly valued, such as business acumen, regulatory knowledge, and customer-centric thinking. Academics can stand out by demonstrating how their skills translate into these areas. **For example, an** academic with experience in pharmacology can highlight their ability to understand drug mechanisms, which is crucial for roles in pharmaceutical development and liaising with regulatory bodies.
- Furthermore, academics should showcase their soft skills, which are critical in industry settings. Skills like teamwork, communication, and working under tight deadlines are essential. **Example:** An academic can describe their role in a multi-disciplinary team that collaborated across departments to publish a significant study, emphasizing their communication skills and ability to integrate feedback from diverse stakeholders, which mirrors the collaborative environment in many biotech companies.
- **Stages of the Recruitment Process for Industry Positions:** The recruitment process in the life sciences industry typically involves several stages, starting with resume screening, followed by phone or video interviews, and finally, in-person interviews that may include a technical assessment or a presentation. **Example:** Candidates can prepare by understanding the specific competencies each stage assesses, such as technical knowledge in initial screenings and interpersonal skills during in-person interviews.
- Academics should also prepare for behavioral interview questions common in later stages. These questions help employers assess how candidates handle specific situations based on their past experiences. **Example:** Preparing the STAR method (Situation, Task, Action, Result) to answer questions can help candidates effectively convey their experiences and showcase how their background prepares them for the challenges of an industry role.
- **Demonstrating Transition Capabilities from Research to Business:** Academics transitioning to industry must demonstrate their ability to apply research skills in a business context. This includes understanding how scientific research can drive product development, influence business strategies, or lead to operational improvements. **For example, an** academic could discuss how their research on molecular pathways has potential applications in developing targeted therapies, showcasing their ability to link scientific discovery to product innovation.
- They should also highlight experiences with direct or indirect exposure to commercial activities, such as collaborating with industry partners,

engaging in technology transfer, or participating in commercialization panels. **Example:** Describing involvement in a project that transitioned from the lab to a startup company can illustrate an understanding of the business lifecycle and the various challenges encountered.

- **Effective Networking Strategies:** Networking is crucial for academics looking to enter the industry. This can be facilitated through professional organizations, LinkedIn, and conferences focused on industry trends. **Example:** Joining groups such as the American Association for the Advancement of Science (AAAS) or attending industry-specific events like BIO International can provide networking opportunities and insights into industry expectations.
- Academics should also consider informational interviews with industry professionals. These discussions can provide insights into what life in the industry is like and what skills and experiences are most valuable. **For example, an** informational interview with a product manager in a biotech firm can reveal the daily responsibilities of the role and the industry's culture, helping academics tailor their applications to fit better the profile sought by employers.
- **Common Misconceptions About the Industry:** One common misconception among academics is that industry roles are less intellectually stimulating than academic research. It is crucial to dispel this myth by highlighting the dynamic challenges and innovations inherent in industry roles. **For example, an** academic transitioning to a role in drug development might find themselves directly involved in designing clinical trials or in strategic decisions about which compounds to push forward, which involves high-level scientific analysis and creative problem-solving.
- Another misconception is that industry positions are profit-driven and do not value the rigorous scientific method. It can be shown that many companies in the life sciences sector are deeply committed to science and innovation as the core of their business models. **Example:** Employees in biotech firms often engage in cutting-edge research published in top journals and collaborate extensively with academic institutions, thereby maintaining a solid connection to the academic community and its values.
- Examples of Successful Transitions: Highlighting successful transitions from academia to industry can provide practical insights and inspiration. Consider a former academic researcher who became a clinical scientist at a pharmaceutical company. By leveraging their deep knowledge of oncology, they contributed to developing new cancer treatments, showcasing the potential impact of academic expertise in a commercial setting. This

transition highlights the applicability of scholarly research in industry and the broader impact on patient care.

- Another example involves an academic who transitioned into a role as a regulatory affairs specialist. Using their detailed understanding of biochemistry and molecular biology, they helped navigate complex regulatory landscapes, ensuring that new medical devices and drugs met safety standards before reaching the market. This shift demonstrates how academics can use their expertise to bridge the gap between innovation and market release, ensuring that groundbreaking discoveries translate into accessible healthcare solutions.
- **Typical Interview Questions for Industry Positions:** Academics should be prepared to answer interview questions that probe their technical competencies and adaptability to the fast-paced industry environment. **Example:** Questions might include, "Can you describe a project where you had to learn a new technology or methodology quickly?" This tests an applicant's ability to adapt, a crucial trait in industry settings where technology evolves rapidly.
- Another common question focuses on teamwork and collaboration, such as, "Tell us about a time you worked with a team under a tight deadline." **Example:** Academics can use such questions to discuss interdisciplinary research projects or collaborative grants, highlighting their capability to work effectively with diverse teams and manage stress, aligning their experience with industry demands for teamwork and efficiency.
- **Career Development Opportunities:** Once in the industry, academics should actively pursue continuous learning and professional development opportunities to enhance their career trajectory. **Example:** Many companies offer training programs, sponsor advanced courses, or support certifications in areas like project management or regulatory affairs, which can be crucial for career advancement.

6

Testimonials

Abstract This chapter features testimonials from individuals who have successfully transitioned from academia to various roles in the life sciences industry. These personal accounts provide insights into the challenges and rewards of career transitions, highlighting the diverse paths available and the transferable skills that are valued in the industry. The testimonials offer valuable advice and encouragement to those considering a similar transition, showcasing the potential for personal and professional growth beyond academia.

Michele Marass, Scientific Advisor—Publication Strategist

Michele Marass

The original version of the chapter has been revised. A correction to this chapter can be found at https://doi.org/10.1007/978-3-031-73565-3_8

© The Author(s), under exclusive license to Springer Nature Switzerland AG 2024, corrected publication 2024

S. Gaburro, E. Rieux, *Career Options in the Life Sciences: Guide Research for Your Path Beyond Academia*, https://doi.org/10.1007/978-3-031-73565-3_6

- **What is your educational background? (initial training, PhD, postdoc, etc.)**

  I began my academic journey with a Bachelor's and a Master's in Industrial Biotechnology at the University of Parma. My curiosity and passion for science took me on an Erasmus exchange to Germany, which was a pivotal experience for me. I then pursued a PhD in Cardiovascular Development at the MPI of Molecular Organ Biology, working on hematopoiesis, angiogenesis, and heart regeneration in zebrafish. Unlike many of my peers, I decided against a postdoctoral path. Instead, I ventured directly into the clinical field within the editorial realm, first with Springer Nature and later as an Editorial Consultant at the MPI of Molecular Cell Biology and Genetics.

- **During your studies, what was your intended career?**

  Initially, my ambition was to merge industrial biotechnology with my field of study, especially during my internship abroad, where I shifted my focus toward Cardiovascular Biology. Early in my PhD, I aspired to an academic career, envisioning myself as a professor. However, as I delved deeper, I realized that my true passion lies in exploring a variety of projects rather than specializing and focusing on one topic, which seemed necessary for a successful postdoc and junior group leader position. This realization led me to seek broader engagements within the scientific community.

- **Did you work in academia for a bit? What prompted you to take a job outside of academia?**

  Yes, I spent some time in academia during my PhD. However, I was keen on diversifying my experiences and impact, which led me to explore opportunities beyond the traditional academic route. The decision was driven by my desire to be involved in multiple projects and to make a broader contribution to science, which I felt was somewhat limited by the singular focus of postdoctoral research.

- **When you look back, do you feel you had more opportunities outside of academia?**

  Looking back, stepping outside academia opened up a wealth of opportunities for me. It allowed me to grow personally and professionally, engage in a variety of projects, and achieve a level of stability and satisfaction that I found challenging to attain within the confines of academia.

- **Which qualities do you think are needed for the role you've had?**

  Decision-making, focus, time management, and creative thinking have been crucial in my roles as an Editorial Consultant and in scientific advising. Balancing personal inclinations with professional decisions, especially in the context of scientific publishing and consulting, requires a nuanced understanding of the field and the ability to make informed judgments. My drive to explore various scientific topics has fueled my career, emphasizing the importance of adaptability and a broad perspective on science.

- **How has your transition been beneficial in terms of personal development, project management, and income?**

  This transition has been immensely beneficial. It has fostered my development in both hard and soft skills, improved my project management capabilities, and provided me with a stable income, especially as a scientific editor with a permanent position. The move to a more diverse scientific engagement has also enhanced my ability to manage priorities effectively and to collaborate across a wider spectrum of the scientific community and several teams within a bigger corporation when dealing, for instance, with resources and regulations.

- **What words of advice do you have for young scientist graduates?**

  My advice to young scientists is to remain open to all career paths. Embrace your entrepreneurial spirit by exploring PhD opportunities in industry settings and by seeking early exposure to practical applications of your research. This approach will not only broaden your career prospects but also enhance your adaptability and understanding of the diverse roles that science can play in society and the economy.

Torben Hager, Product- and Market-Manager

Torben Hager

Before

Account Executive Germany at Data Sciences International/Harvard Bioscience (4× years)
Chief Scientific Officer at Biobserve GmbH (4× years)
2× Postdocs at CNCR Amsterdam/Sylics B.V.
PhD student/early-stage researcher at VU Amsterdam/Marie Curie ITN "Brain Train"

Questions

- **What is your educational background? (initial training, PhD, postdoc, etc.)**

My diploma in Neurobiology and Zoology I received from the Ruhr-University Bochum. Afterward, I began my PhD studies at the Vrije Universiteit Amsterdam in the Behavioral Neuroscience Research group of Dr. Oliver Stiedl (within the Department of Functional Genomics headed by Prof. Matthijs Verhage and the CNCR Amsterdam). My research focused on novel approaches to enable holistic phenotyping of emotional behavior in mice with increased validities and significance. My PhD project was part of the Marie Curie Initial Training Network (ITN) called "Brain Train" with 15× academic and 3× industrial partners all around the globe, with a strong focus on a comprehensive, interdisciplinary range ("from bench to bedside"). My PhD thesis is titled "Advanced autonomic and behavioral phenotyping of emotional behavior of mice" and the PDF can be accessed for free at the library of the VU Amsterdam (link to thesis: https://research.vu.nl/en/publications/advanced-autonomic-and-behavioral-phenotyping-of-emotional-behavi). Following my PhD defense, I spent 2x years with Postdoctoral research at the CNCR Amsterdam, working on the optimization, validation, and commercialization of my scientific approaches.

- **During your studies, what was your intended career?**

Initially, before and during my diploma at the RUB in Bochum, I intended to pursue a career in academic research. Importantly, especially during my first years in academic research, my desired focus (i.e., specialization within biomedical research) changed basically every 6x month on average. Each course, practical, lecture, and experience provided me with novel perspectives I was not aware of in their wonderful details. Only during my PhD (and especially due to the wide and interdisciplinary team within my Marie Curie network) I slowly began shifting my general expectations and evaluating opportunities beyond the *academic ivory tower*. Specifically, I learned about alternative perspectives beyond my previously rather limited horizon through the tight connection with the several industry partners. One of these industry partners employed me after my 2x Postdocs in Amsterdam and gave me the opportunity to be an active part of biomedical research beyond academia.

- **Did you work in academia for a bit? What prompted you to take a job outside of academia?**

Yes, I did (1× PhD and 2× Postdocs). Importantly, I enjoyed this time endlessly and I still share the *romantic* idea of serving a higher purpose for global human health and flourishing in intellectual freedom. However, the more I became aware of earthly necessities with increasing age (and responsibility outside the ivory tower), neglecting important perspectives such as personal and financial needs for the sake of academic freedom seemed more and more shortsighted to me. Being directly involved in the consequences of the concept of "publish or perish," my courage was lost to my rational self, and I left academic research.

- **When you look back, do you feel you had more opportunities outside of academia?**

  Yes, surely. This realization set in only at the end of academic career and is still growing every day. While I was increasingly aware of opportunities beyond research itself also in academia early on in my career (i.e., technology transfer, animal welfare, grant, and budget justification) I never felt challenged enough by what seemed to be a rather limited horizon in my eyes. Only by looking at the vast field of opportunities outside of academia, I began to realize how challenging opportunities in administrative positions, sales, product and project management (and much more) within my most favorite biomedical research sector can be.

- **Think in terms of personal development (hard and soft skills), opportunities to manage projects/budgets, personal income, etc.**

  In my opinion, the by far most important skill set is based on human skills such as communication and networking skills, emotional intelligence, and empathy. Of course, organizational, logical, and scientific skills are very important as well, but will not bring you far, as soon as the first person you are collaborating with hates to communicate with you.

- **Which qualities do you think are needed for the roles you've had?**

  As indicated above, I strongly believe that the most essential soft skills are communicating and networking. Only my contacts around the entire community within (biomedical) research enabled me to look beyond my self-inflicted tunnel vision and ultimately actively helped me to get access to those challenging opportunities I was/I am so happy to be involved with. A strong example here: Two of the industry positions during my career were mediated and finally initiated by a former scientific connection that morphed into my best friend over the years.

- **What words of advice do you have for young scientist graduates?**

  Leave your comfort zone and forget about your (or your friends) prejudices. Only looking beyond your self-inflicted ivory tower/tunnel vision will give you comprehensive insights you can base your decisions on.

Frederick Odun-Ayo, Product Manager

Frederick Odun-Ayo

- **What is your educational background?**

  I had my honors degree in Microbiology from the University of Ilorin, Nigeria and graduated top three in my master's degree in Virology from the reputable University of Ibadan, Nigeria. I obtained my doctorate in Biotechnology from Durban University of Technology, South Africa where I was awarded a scholarship for my doctorate research. I progressed my training as a Postdoctoral Research Fellow to acquire different research skills including mouse cancer models and advanced microscopy using state-of-the-art facilities within the University of KwaZulu Natal Biomedical Research Unit and Optics Imaging Centre, one of the largest laboratory animal facilities and microscopy facilities, respectively, in Africa. Later, I joined the biotechnology research team as a postdoctoral research fellow at Cape Peninsula University of Technology, South Africa, where I explored medicinal plants and microbial products to develop an immune checkpoint inhibitor in colorectal cancer.

- **During your studies, what was your intended career?**

  While pursuing my studies, I aspired to become a virologist and specialize in medical biotechnology with the ultimate goal of achieving a professorship.

- **Did you work in academia for a bit? What prompted you to take a job outside of academia?**

I have 7 years of experience in research and teaching at various universities. With an impressive track record of 20 publications in reputable international journals and over 100 citations, I have taught and trained both undergraduate and postgraduate students. However, there are several reasons why I decided to take a job outside of academia. Firstly, securing a permanent lecturer position is challenging and slim, as promotion opportunities are limited, and new positions are not often created. I spent years in postdoctoral training, but the job insecurity and uncertainty became unbearable. One never knows if their contract will be renewed after a year, and it depends heavily on fund availability for continued sponsorship. I noticed that my peers working in the industry are building successful careers and receiving better opportunities, which outweigh what I could get in academia.

- **When you look back, do you feel you had more opportunities outside of academia?**

I enjoy working in the fields of research and education. My current role allows me to collaborate with researchers and scientists from various universities and assist them in achieving their research objectives and working efficiently. I also provide training and help them develop their skills, a form of teaching that I find fulfilling. Additionally, I can attend various conferences, symposiums, and forums that are not easily accessible or affordable in academia.

- **Think in terms of personal development (hard and soft skills), opportunities to manage projects/budgets, personal income, etc.**

Apart from the benefits I mentioned earlier, working in the industry also helps me develop important skills such as creativity, critical thinking, and proactivity. Additionally, it puts you in a position to adapt to a fast-paced working environment. It exposes you to new technology, environments, people, and systems, which has contributed significantly to my personal growth. Of course, job security and a good income with added perks are guaranteed. In terms of project and budget management, while in academia you may have more control, the trend is now moving toward collaborative studies where decisions are made jointly, and not just by one person.

- **Which qualities do you think are needed for the role you have had?**

  Among many skills that are essential for success include the ability to work independently, meet deadlines, communicate effectively both verbally and in writing, skills in grant writing, networking, and the ability to collaborate when necessary.

- **What words of advice do you have for young scientist graduates?**

  I would like to say they should keep their options open and not limit themselves to the walls of academic careers but rather explore opportunities to thrive in different areas.

Barbara Mascialino, Associate Director—Senior Scientist in Pharmaceutical industry

Barbara Mascialino

- **What is your educational background? (initial training, PhD, postdoc, etc.), what is your current role/job/profession?**

  Masters in nuclear physics at University of Genova, Italy; PhD in Sleep medicine/biostatics, Postdoc on Epidemiology and Biostatistics at Karolinska Institute Stockholm, Sweden. I am currently an Associate Director—Senior Scientist at GSK. My role is to handle real-world evidence for clinical studies in oncology, contributing to market access efforts for novel drugs. I have also been working with the European Commission where my role evolved from evaluating proposals to supervising and managing experts, and eventually becoming an independent observer for the health calls of Horizon Europe.

- **During your studies, what was your intended career?**

  I had a strong interest in becoming a medical doctor but realized I was very much attracted to the theoretical side of medicine. So I shifted my studies to medical applications of physics and other disciplines such as statistics.

- **Did you work in academia for a bit? What prompted you to take a job outside of academia?**

  I worked in academia for a few years during postdoc positions. Several reasons pushed me to look into industry. First is some form of internal competition, for example, between physicists and biologists focusing on their respective fields and publications, rather than looking for ways to best collaborate. Second were personal reasons, having to care for my mother and the arrival of my first child. Finally, the challenges associated with constant grant applications in academia to fund one's research.

- **When you look back, do you feel you had more opportunities outside of academia? Think in terms of personal development (hard and soft skills), opportunities to manage projects/budgets, personal income, etc.**

  My first role as a scientist was with a Contract Research Organization (CRO), and I have to say I was a bit cautious at first. However, I realized my work as a scientist was very interesting and had a direct impact on developing novel therapies for patients. And I had access to resources available from established budgets. I also enjoyed the higher tempo of the industry, which contrasted with the slower pace of academia. I later joined a life sciences company where I contributed to health economics, developing models and evidence to facilitate market access for diagnostic tests. And of course, working in industry provides a stable source of income, which is especially important as one starts having a family. I am also grateful for contributing to European research and working with cross-functional teams, this has been very enriching from a personal perspective.

- **Which qualities do you think are needed for the role(s) you have had?**

  Resilience, adaptability, effective communication in cross-disciplinary settings, and a passion for making a meaningful impact on healthcare.

- **What words of advice do you have for young science graduates?**

  I encourage young graduates to be open to new opportunities in new places, to be curious, and to focus on effective communication skills. I cannot stress enough the importance of simplicity and clarity in communication, and the ability to summarize complex ideas for broader understanding.

Jean-Pierre Valentin, Senior Scientist and Director in Pharmaceutical Industry

Jean-Pierre Valentin

- **What is your educational background? (initial training, PhD, postdoc, etc.), what is your current role/job/profession?**

PhD in Cell Biology and Physiology from the University of Montpellier, France, and Postdoc at the University of California San Francisco in Cardiorenal physiology and Pharmacology. I am currently Senior Director and Head of Investigative Toxicology at the UCB Pharma in Belgium.

- **During your studies, what was your intended career?**

I would like to give a bit of background to explain my career path. In my teenage years, I was strongly influenced by two individuals: one was a medicinal chemist working on projects in Africa and the other was working on mapping ocean floors. Both were traveling the world as scientists, which I found very inspiring. Then I became truly fascinated with medical research during my university days. I worked on natriuretic peptides to treat hypertension which was quickly emerging and impacting many, then on novel approaches like monoclonal antibodies. After my Postdoc in the USA, I was naturally looking for positions in Academia back in France. I also applied for positions in the pharmaceutical industry, and have to say that I knew very little about it then.

- **Did you work in academia for a bit? What prompted you to take a job outside of academia?**

Apart from my Postdoc position, I never worked directly in Academia. My first industry position was with Pierre Fabre Research Centre in France, I discovered the industry's many opportunities, available resources, and the ability to focus on impactful science. This contrasts with the constant struggles academics face for funding and the "publish or perish" conundrum.

- **When you look back, do you feel you had more opportunities outside of academia? Think in terms of personal development (hard and soft skills), opportunities to manage projects/budgets, personal income, etc.**

As mentioned earlier, not having to worry, or spending time securing funding from grants allowed me to focus on the research, and the people around me. In industry, I had a lot of autonomy and continuous learning opportunities despite the lack of absolute project control, since projects are aligned with company's goals. I learned to develop many "people skills", communication, and teambuilding. This is not taught in universities, or at least it was not taught then. Working in safety assessment, that is prior to the first clinical tests in humans, I got to support many different research portfolios;

this is fantastic because you work with and learn about so many different therapeutic areas. I recall working on a drug that was approved and became a medicine, and I found out later that older members of my family were taking that drug and it was making a difference in their life. That is a priceless reward and a strong motivation to continue.

- **Which qualities do you think are needed for the role(s) you have had?**

  I think one needs to balance scientific excellence with pragmatism by aligning good science with industry goals. I often use the 80/20 rule, focusing on efforts that yield significant responses (80%) with minimal effort (20%) rather than striving for near-perfect outcomes, which demand excessive effort. Understanding these concepts is crucial for achieving success while making life easier for the team. I think that the team's overall success is determined by the weakest link. I think that surrounding oneself with talented individuals, being able to recruit, retain, and nurture them is a critical quality for success.

- **What words of advice do you have for young science graduates?**

  I would encourage graduates to explore diverse opportunities, keep an open mind, not shy away from seeking guidance, and pursue what they genuinely enjoy while being open to new experiences and connections. If I had known more about the dynamics of industry during my studies, I may have tried to pursue a PhD in industry. But no regrets on anything, all experiences are useful and learning opportunities.

Sebastien Moussay, Founder and President

Sebastien Moussay

- **What is your educational background and what is your current role/job/profession?**

  I first started with a 2-year program to become a sales representative but quickly realized I could not sell things that people did not need! I had a strong interest in sports, so I started over in Life Sciences and Sports Sciences at Paris V University, and then a PhD in Sports Sciences, at Caen-Normandy University.
  I am now the president of Bodycap, a company I co-founded in 2011.

- **During your studies, what was your intended career?**

  I was aiming for a career in Academica, combining teaching and doing research at the university level.

- **Did you work in academia for a bit? What prompted you to take a job outside of academia?**

  At first, my research work was purely academic. I then became really interested in the development of technologies that would serve scientific purposes. In 2016, I started a collaboration with an industrial company to focus on innovative ways of collecting physiological data in an ecological environment. That led me to combine my university position with a role in the new company we started. In 2014, I took a leave from my university position to become an officer of the company.

- **When you look back, do you feel you had more opportunities outside of academia? Think in terms of personal development (hard and soft skills), opportunities to manage projects/budgets, personal income, etc.**

  There is a world of differences between academic and entrepreneurial environments, especially with small business companies. With a university-tenured position, there is job security and a relatively low level of stress, but often administrative complexity. On the other hand, in the corporate world, there is no "for life" job security and the need to make quick decisions. My experience starting a new business has allowed me to develop many skills in planning, finance, human resources, management, recruitment, and effective communication.

I am also very grateful for meeting so many fantastic people who helped and supported me when we started the business. And now given the nature of our products, I often have the privilege to meet Olympic or extreme sports athletes, this is very rewarding!

- **Which qualities do you think are needed for the role(s) you have had?**

  I think one must be agile and adaptable. When you start a business, many unknown challenges will pop up and you must deal with them. As a business owner or partner, there are no benefits such as paid vacation and unemployment insurance so, one must be self-confident and resilient. I learned to be responsible for my actions, for my 15-person team, and in general for the long term of the business. For example, I am very proud that our business has a super low level of absenteeism compared to the national average.

- **What words of advice do you have for young science graduates?**

  I encourage young people not to fear failure and not to fear leaving a path that we think we have to follow because of our education. I think French culture still stigmatizes failure, unlike other cultures where failure is seen as a learning opportunity. I encourage you to take risks, to try and try again. There are so many opportunities to bring value to many enterprises, learn, and develop on a personal level.

Yanglei Zhang, Application and Technical Support Specialist

Yanglei Zhang

- **What is your educational background? (initial training, PhD, postdoc, etc.), what is your current role/job/profession?**

  I graduated with a degree in veterinary medicine and pursued a master's focusing on Preventive Veterinary Medicine with an emphasis on immunology and molecular biology. Presently, I serve as the sales director for a clinical contract research organization (CRO) in China, specializing in software for digitalizing clinical trials.

- **During your studies, what was your intended career? Did you work in academia for a period of time? What prompted you to take a job outside of academia?**

  I was split between pursuing a research path in a laboratory and entering the software or instrument industry. At the time the job market was a bit tough in China for both Academic and industry positions. Eventually, I landed a role with a US research instruments company as an application support specialist. This turned out to be my dream job: I got to work and collaborate with many different labs in both Academia and Industry, I got to travel inside and outside China. It also satisfied my scientific interests as I helped researchers in many different fields like cardiovascular, neuroscience, and respiratory.

- **When you look back at your career, do you feel you had more opportunities outside of academia?**

  My first position as an Application and Support specialist allowed me to better understand possible opportunities in the industry. I am glad I chose that path as it has allowed me to grow by building a set of skills during each professional experience. I later transitioned from technical support to managerial roles. Adapting to diverse regulatory environments and industry-specific nuances became a crucial part of my journey.

- **Which qualities do you think are needed for the role(s) you have had?**

  I think the ability to listen and understand the scientific and logistical needs of researchers is very important. My first year was very intense as I had to learn and understand so many things. Troubleshooting and resolving technical issues requires patience, perseverance, resilience, and good communication skills. When customers call, it is because they have a problem that needs to be resolved, often with a fast timeline. I often had to figure out things on my own but with the help of a team, so teamwork is essential too. Finally, one must be able to work following strict documented processes with clients working in highly regulated medical environments.

- **What words of advice do you have for young science graduates unsure about their career path?**

Reflecting on my career path, I realize the importance of aligning personal interests with professional aspirations. My advice to young science graduates would be to seize opportunities, persevere once a pathway is chosen, and avoid dwelling on regrets about unexplored options. Continuous learning has been my mantra, I think that progress should be considered personal not comparative to others. I think everyone should embrace a continuous learning mindset, it will lead to personal and professional development.

Eric Delpy, Scientific Director in Contract Research Organization (CRO)

Eric Delpy

- **What is your educational background, and what is your current role/job/profession?**

  I am a PhD and postdoc in applied Cardiovascular Pharmacology. I am currently Preclinical Scientific Director at Biotrial, a French and US based CRO.

- **During your studies, what was your intended career, did you work in academia for a period of time? What prompted you to take a job outside of academia?**

  After completing my master's degree in physiology with a focus on pharmacology, I continued with a research master's in Paris, the practical part of which was carried out at CNRS, specifically diving into electrophysiology and the action of drugs on ion channels in neurology. There, I began to see the more practical aspects of applied research and understand the link between basic pharmacology and drug development.
  My original goal was to pursue a doctoral thesis in the same lab within Academia; however, I did not secure a public grant as options were very limited. I therefore seized an opportunity to do my PhD in the pharmaceutical industry at GSK, working on PDE5 inhibitors initially developed in the cardiovascular domain. Looking back, my interests were definitely in applied over fundamental research, so if I had had the choice, I would still have chosen a PhD in the industry. I then did a postdoc within Academia at INSERM Bichat Hospital in Paris, which completed my hands-on training and finally, further confirmed my interest in applied research. I then had the opportunity to join Biotrial where I stayed for 15 years: I started as a cardiovascular and respiratory Project Leader and ended as Director of Preclinical Operations. After this long period of loyalty and many other opportunities, I joined a Biotech company for 4 years, which allowed me to discover the field of medical devices, another strategy for developing innovative therapies for patients. I recently had the opportunity to join Biotrial again, this time as Preclinical Scientific Director.

- **When you look back at your career, do you feel you had more opportunities outside of academia? Think in terms of personal development (hard and soft skills), opportunities to manage projects/budgets, personal income, etc.**

  The CRO world is very different from Academia and even from the pharma industry, mainly due to the existence of a research contract and the customer relationship. CROs are typically environments where multitasking rules. I managed studies and research projects, teams, and budgets, while being involved in commercial and regulatory aspects. This demands flexibility, constant adaptation, and the ability to handle stress and juggle multiple responsibilities. I found myself traveling extensively, an incredibly enriching experience and opportunity to meet people and expand knowledge and network.

- **Which qualities do you think are needed for the role(s) you have had?**

  My years in CRO and biotech companies allowed me to develop diverse skills, ranging from time management to handling crises under pressure. I often had to step out of my comfort zone, deal with complex situations, and manage tensions when necessary. However, each challenge was a lesson, a way to learn, react, and adapt, all those situations that we can also experience in our personal lives. I also had to learn when to say "no" in diplomatic terms and establish clear ethical boundaries, especially crucial in the scientific field where pressure can sometimes influence decisions. Finally, as a team manager, I learned empathy, necessary kindness, and above all, the importance of active listening.

- **What words of advice would you have for young science graduates unsure about their career path?**

  My advice would be to dare to explore different paths without fearing change or the unknown. Professional choices are not definitive, so being flexible and open is essential to achieving personal development in an ever more dynamic workplace.
  I also think knowing oneself well is key, and one should always remain true to one's own values and principles. It is also important to develop a professional network of people who can help and guide us.

Elena Hoffer, Co-Founder Alma.me

Elena Hoffer

A typical day in my life as a Co-Founder and CEO of Alma.me looks like this: As we are in the early phase of Alma.me, I am responsible for everything. Typically, I create content for LinkedIn and other social media channels, have meetings with my Co-Founder (daily), and engage with prospective customers. I make strategic decisions on how to move forward and prepare for talks and presentations.

- **What is your educational background? (initial training, PhD, postdoc, etc.), what is your current role/job/profession?**

  I trained to become a pharmaceutical technical assistant to work in the pharmacy. After completing my training, I began working in the pharmaceutical industry for 1 year. I then decided to pursue a bachelor's and master's degree in pharmaceutical biotechnology at the University of Applied Sciences in Biberach and the University of Ulm in Germany, respectively. Following this, I pursued a PhD in Medical Sciences at the Karolinska Institute in Stockholm, Sweden. Immediately after completing my PhD, I became the CEO and Co-Founder of my own company called Alma.me, which assists PhDs in their transition from academia to industry.

- **During your studies, what was your intended career?**

  During my bachelor's and master's degrees, I considered working in the pharmaceutical industry, but I was not sure about the specific role I wanted. It was during my master's thesis that I became interested in pursuing a PhD in the field of immunology.

- **Did you work in academia for a bit? What prompted you to take a job outside of academia?**

  I worked in academia as a research assistant and PhD student for a total of 7 years. I chose a job outside of academia because I wanted to feel more valued and make a living. Additionally, I wanted to make an impact with my own company and create something myself.

- **When you look back, do you feel you had more opportunities outside of academia?**

  As a general answer, I think staying in academia or leaving is not only about opportunities but also about how one envisions their life: work-life balance, salary, professional development, responsibility, etc.

- **Which qualities do you think are needed for the role(s) you have had?**

  Your current role as a Co-Founder and CEO requires:

- Persistence and resilience
- Problem-solving abilities
- Creativity
- Self-motivated working with little to no structure
- Critical thinking
- Time and project management
- Communication skills (written and oral)
- Presentation skills
- Analytical skills
- Adaptability
- Being comfortable getting out of your comfort zone
- A "get stuff done" mentality

- **What words of advice do you have for young science graduates?**

  Start building a network early, whether you plan to stay in academia or transition to another field. A robust network can open doors and provide opportunities, particularly in an industry where many positions are filled through networking.
  Expand your horizons beyond your academic focus by getting involved in organizations or activities outside of your field of study. This broader experience can enrich your skill set and enhance your professional development.

Cindy Leyton, Product Manager at Regeneron

Cindy Leyton

- **What is your educational background? (initial training, PhD, postdoc, etc.), what is your current role/job/profession?**

  I completed my BS in Biochemistry at the University of New Haven and my PhD in Pharmacology from Stony Brook University (SBU). My first postdoc at Stony Brook Medicine was in the Department of Anesthesiology where I studied neuroscience, and my second postdoc was in the Department of Pathology where I studied cancer biology, therapeutic targeting, and biomarker development.

- **During your studies, what was your intended career?**

  I always wanted to be a professor as I loved teaching and helping others grow in their career, especially in science. I was awarded several special grants and fellowships dedicated to empower women and minorities in sciences to become faculty, and I was very dedicated and set on becoming a professor until it finally happened after my second postdoc.

- **Did you work in academia for a bit? What prompted you to take a job outside of academia?**

  I became a non-tenure track Research Assistant Professor after my second postdoc in the same lab I did my postdoc. As I grew in seniority and expertise in my postdocs, my responsibilities shifted from bench science to scientific/project management more and more. When I became a professor, I was functioning in many ways like a program manager given that I led multiple research programs, taught students, wrote grants, assembled scientific teams, and coordinated experiments and other key research activities. Although I very much loved my job as a professor, I encountered many financial obstacles due to the relatively low pay combined with the high payments associated with student loans. I also became very sick and was not sure where I could continue working to the demands of an academic lifestyle (late nights/weekends), and I also began feeling imposter syndrome. I spent about 2 years doing informational interviews, applying to jobs, learning from failure in applications, and interviews, and eventually, I realized that becoming a program manager in pharma made sense as that was a very direct way to enlist the use of my skill sets and be productive outside of the lab.

- **When you look back, do you feel you had more opportunities outside of academia?**

  I wish I had been able to do an internship or co-op to prepare me further to understand what non-academic jobs are like. This may have allowed me to learn about how science occurs outside of academia in roles that do not require one to be on the bench to be productive in science.

- **Which qualities do you think are needed for the role(s) you have had?**

  Being highly resilient, having social skills, being very diligent about prioritizing, planning, and executing. Being able to build high-performing teams, and creating environments where the teams can get things done with little distraction or "churn."
  Resilience, adaptability, effective communication in cross-disciplinary settings, and a passion for making a meaningful impact on healthcare.

- **What words of advice do you have for young science graduates?**

  Starting exploring career options early in your training, do not wait until the year you want to leave because finding out what you want to do outside of academia is not straightforward and takes time to do career exploration. Learning how to receive feedback and put it to use. Many people get defensive or are not willing to listen when others are giving them feedback. This not only results in little to no growth but also in breaking relationships with those who intend to help you. Learn how to listen to and incorporate feedback.
  Your attitude is something you can control. Your perception of any situation (e.g., whether you get triggered/angered/confused) is in your control. Therefore, practice learning how to control your impulses. Next time you get defensive or angry, etc. you will learn to start positive, patient, and prudent.

Mushtaq Bilal, Postdoc and Entrepreneurs AI for Academic

Mushtaq Bilal

- **What is your educational background? (initial training, PhD, postdoc, etc.)**

  I have a PhD in comparative literature from Binghamton University, New York. Currently, I work as a postdoctoral researcher at the University of Southern Denmark's Hans Christian Andersen Center.

- **During your studies, what was your intended career?**

  Frankly, I had no idea what would happen after my PhD. I applied to a bunch of jobs in academia in various parts of the world but got rejected every time. It is very hard to get a decent job with a humanities PhD and I would never recommend anyone to do a PhD in the humanities. I did get a teaching job at a Pakistani university but I did not want to stay there and moved to Denmark at the end of 2022.

- **Did you work in academia for a bit? What prompted you to take a job outside of academia?**

  I still work in academia. I do not want to switch to a non-academic job as much as start my own business.

- **When you look back, do you feel you had more opportunities outside of academia?**

  Not really. I was born and raised in Pakistan in a working-class family and academia was one of the only ways available for me to become upwardly mobile.

- **Think in terms of personal development (hard and soft skills), opportunities to manage projects/budgets, personal income, etc. what qualities did you need?**

  I would have loved to do a few courses on computer science.

- **Which qualities do you think are needed for the role you have had?**

  I want to say hard work but that is only partially correct. You need to build a strong professional network around you and that takes a lot of time and effort.

- **What words of advice do you have for young scientist graduates?**

Do not do a PhD unless you are insanely rich in which case you are not reading this. Do a PhD if you want to stay poor for most of your life.

Finish your undergraduate degree, learn business, marketing, and sales. Learn skills that put food on the table. Forget about science. It is a rich folks' sport. Once you have a few million sitting in your bank account, then think of doing a PhD. You will enjoy it more and feel the experience is intellectually rewarding.

Diana Roccaro, Clinical Data Manager

Diana Roccaro

- **What is a typical day of yours?**

Since half a year, the moment when I had quit my amazing job as a developer of Artificial Intelligence for customers in the healthcare industry (NLP—natural language processing), I have been trying to follow my heart from morning to evening. Thus, every day has been different. One month ago, I spontaneously decided to book a trip to northern Scandinavia to benefit from this season's last chance to hunt the Northern Lights. I have taken a course aimed at improving relationships (as well as our entire life, to make our deepest dreams come true), these past 10 months. Everything I have done was heart-driven, and my daily tasks were as variable as life is. Since 2 weeks, I am back to on-site work, after 4 years of remote-only work settings. My responsibility as a data manager, at the intersection between academic research, a university hospital, and a tissue bank, is to bring meaningful structure and logic into data collected for research purposes. Since healthy nutrition and lifestyle are of high importance to me, I love starting my day in the kitchen, preparing some tasty breakfast rich in vitamins and proteins. Thereafter, I typically start working on computer screens. Whenever working at home, I tend to interrupt computer sessions by doing physical work in the garden or the necessary housework. I like to spend free days and weekends out in nature and mountains, but also have many private projects that keep me busy on computer screens for many hours.

- **What is your educational background (initial training, PhD, postdoc, etc.)? What is your current role/job/profession?**

After five semesters of human medicine, I had realized to be more interested in science and research than in becoming a physician myself. I switched to study biomedical sciences and graduated with a PhD in medical sciences/neurosciences. In parallel, I had started working as a medical writer, medical translator and proofreader. Both in the domain of regulatory affairs (medical devices) and the academic setting (peer-reviewed publications). My last job as a scientific writer and assistant introduced me to the field of machine learning and data science, which led me to self-learn programming in Python, and to contribute to IT products for hospital customers as a healthcare data scientist in entirely remote working settings. After a "career break" lasting for 5 months, in which I focused entirely on myself and my own business ideas, I am now enjoying my new on-site job as a clinical data manager. In parallel, I am defining the scope of my own "business babies" that I am about to start as an independent entrepreneur.

- **During your studies, what was your intended career?**

I started studying human medicine without having a clear view of what I would do with it 1 day. I had never consciously intended to become a doctor myself, but I had known that there will be many alternative career paths, on top of working as a physician per se. After having switched to biomedical sciences, it became somehow self-evident to continue by myself doing research. Which led me to finally pursue my PhD in Neurosciences.

- **Did you work in academia for a bit? What prompted you to take a job outside of academia?**

After finishing my PhD, I accepted the invitation of my PI to provide continued support to one of his Master's Students, in a part-time engagement at a low workload. In addition, just before the outbreak of the COVID pandemic, I had accepted a second offer from him, which officially lasted for 6 months at 40%, but which we agreed on to distribute over a longer period of time, subject to mutual agreement. Besides these accepted invitations, I have never actively hunted for academic positions myself. My initial transition away from academics to industry was due to the simple fact that I did not find a position in academia that easily, while there were plenty of interesting suitable offers on the industrial side. The medical writer job was the result of a company owner presenting his company and offer during a presentation at my university—he was actively recruiting and hired me immediately thereafter. What matters most to me, is purpose. My academic fundamental science research did not result in a direct application and added value for humans. I like these positions most, through which I can create concrete and immediate value for humans, animals, or this planet, be it academic or industrial, as an employee or self-employed.

- **When you look back, do you feel you had more opportunities outside of academia?**

The offers perceived as my opportunities always depend on where I focus my attention, do not they? I have mainly focused on positions outside academia, but besides others, because my PI got retired shortly after the end of my PhD, and because there was no other professor in my network who would have had a suitable position of interest to me, back then. Regarding my personal needs, I have 2 children (now aged 10 and almost 7), and it was always a prerequisite for me to be able to juggle between private and professional needs, a bit flexibly if possible. I have never liked neither the publication pressure of the academic environment nor the responsibility of having to care about grant writing, and the pressure associated with that. I am a person with very broad interests and talents, who has never stayed in one and the same field for longer than a few years at most. I think that this is the reason why industry positions have appeared more attractive and suited for me, personally, in the past. But this does not mean that I would generally like positions in academia less—I am open to every challenge, as long as it sparks my interest, matches my passions, and allows me some freedom in managing my own time and energy.

- **Which qualities do you think are needed for the role(s) you have had?**

Flexibility to adapt to work environments and collaborators, analytical thinking, clear and concise, audience-adapted interdisciplinary communication, language skills, readiness for life-long learning, patience and endurance, etc.

- **What words of advice do you have for young science graduates?**

Never forget who you are and where your true talents lie. Every position typically requires us to also do one or two things we like less. It is always up to us to choose from the abundant wealth of millions of offers out there. If we believe our options to be limited, we likely tend to stay within our comfort zones and hesitate to dream really big. Everything is possible, if we believe in it and are truly ready to make it happen. The world has grown small, with the possibilities of remote working and the advances of digitalization and the fourth industrial revolution. It will always remain our own responsibility to remember, who we really are and what we like and dislike. We need to take a previously unknown path from time to time, to find those we like more than others. To then re-evaluate whether to stay on our current track, or also once in a while to just take a break for ourselves to find back our truest inner self.

Bruno De Carvalho, Stakeholder Engagement Manager, US Pharmacopeia

Bruno De Carvalho

- **What is your educational background, and what is your current role?**

  I am a specialist in molecular biology and cancer research. I obtained my Master's degree in Environmental Health Sciences with a focus on toxicology (SUNY Albany) and my PhD in Molecular and Cellular Pharmacology (Stony Brook University). Currently, I am the Stakeholder Engagement Manager at the United States Pharmacopeia (USP). I am responsible for training scientists and stakeholders, organizing conferences, workshops, and other activities to promote engagement, and positioning our organization in the biologics sector.

- **During your studies, what was your intended career?**

  During grad school, I realized I loved science but did not necessarily want to be the one doing the research. I was more interested in using my scientific knowledge to communicate science or make business decisions. Therefore, I set my sights on science communication and business-oriented roles within science, such as a Field Application Scientist or a role in Management Consulting.

- **Did you work in academia for a bit? What prompted you to take a job outside of academia?**

  Yes, my work in academia was primarily during my graduate studies, but my interest in industry positions started even before that. I was an undergraduate intern at the pharmaceutical company Abbvie, and this experience opened my eyes. I witnessed professionals using their scientific knowledge to solve problems without necessarily being the ones performing the experiments, and it motivated me to pursue careers where I would use my expertise beyond bench research.

- **Looking back, do you feel you had more opportunities outside of academia?**

  Absolutely. The non-academic roles allowed me to explore a variety of career paths that utilized my scientific expertise in dynamic ways, from science communication to stakeholder engagement. These roles often required a blend of technical knowledge and communication skills, which suited my strengths and interests better than traditional academic research.

- **Which qualities do you think are needed for the role(s) you have had?**

  Effective communication, interpersonal skills, strategic thinking, and a solid understanding of science are crucial. In addition to mastering these skills, it is important to prove to an interviewer that you have them. Therefore, engaging in extracurricular activities will provide you with the confidence to include these words on your resume.

- **What words of advice do you have for young science graduates?**

  My main advice is to engage in extracurricular activities. If you do not know what careers to pursue, extracurricular activities will help you understand more about these career paths and figure out what your passions are. If you already know what your dream job is, extracurricular activities will help you to be prepared for it. In my case, I considered pursuing a career as a Management Consultant and, as a result, I started a consulting club at my university where I studied business cases and provided consulting pro bono to a biotech company. I also wanted to be a Field Application Scientist and ended up creating a YouTube channel to explain scientific methods and consequently demonstrate my communication skills. This channel helped me land my first job after my PhD. My story proves that extracurricular activities are effective, and I advocate that every science graduate should engage in them.

Annette Klaus, Veterinarian and Senior Publishing Editor at Springer Nature Group

Annette Klaus

- **What is your educational background? (initial training, PhD, postdoc, etc.)**

  My educational journey began with 1 year of media studies, followed by a degree in veterinary medicine. I further specialized with a doctoral thesis in wildlife conservation medicine and parasitology. My academic involvement continued as I expanded into laboratory animal science and took up university lecturing, allowing me to blend my passion for animal health with academic pursuits. Most recently, I trained part-time in the area of Education and Science Management.

- **What is your current role/job/profession?**

  I currently serve as a publishing editor at Springer Nature. In this role, I focus on overseeing the publication process for our book portfolio that contributes to advances in veterinary medicine and science, animal welfare, parasitology, and related fields.

- **During your studies, what was your intended career?**

  During my academic studies, I was deeply committed to pursuing a career in wildlife conservation and medicine. I was driven by a passion to apply veterinary science to conserve wildlife and ensure the sustainability of natural habitats.

- **Did you work in academia for a bit? What prompted you to take a job outside of academia?**

  Yes, I spent a significant portion of my early career in academia, engaging in research, teaching, and laboratory animal care. The decision to move outside academia was driven by the limitations posed by time-bound contracts and the general uncertainty of long-term academic positions in Germany. This prompted me to explore more stable and diverse career opportunities in the industry.

- **When you look back, do you feel you had more opportunities outside of academia?**

  Reflecting on my career, I definitely see that stepping out of the academic environment opened up a broader spectrum of opportunities. The move allowed me to apply my skills in new and dynamic contexts, particularly in the publishing industry, where I could influence the dissemination of scientific knowledge more broadly.

- **Which qualities do you think are needed for the role(s) you have had?**

The roles I have held, both in academia and in industry, required a strong set of qualities including adaptability, resilience, time management, and effective communication skills, especially in cross-disciplinary settings. A deep passion for making a meaningful impact in the healthcare and conservation fields has also been crucial. These qualities have enabled me to navigate challenges and contribute effectively across different aspects of my work.

- **What words of advice do you have for young science graduates?**

For young graduates starting out in science, my advice would be to focus on effective communication, time management, and project management skills. Embrace new technologies and stay flexible in your job search. It is also vital to network extensively and follow your interests passionately, as these elements are key in shaping a fulfilling career.

7

Integrating Such Information in a Course at a University for a Master's or PhD's Program in Life Sciences

Abstract This chapter discusses the importance of integrating career development education into academic programs to prepare students for diverse career paths. It proposes a modular, credit-bearing elective course that incorporates workshops, mentoring, and practical job exploration tasks. The chapter also addresses the challenges of dispelling misconceptions about non-academic careers and emphasizes the need for continuous updates and industry partnerships to ensure the course's relevance and effectiveness. It concludes by outlining key performance indicators (KPIs) to assess the course's impact on students' career preparedness and success.

A strategic approach is necessary to integrate a course that broadens career pathways for master's and PhD students, aligning with existing academic frameworks and catering to the growing demand for diverse professional opportunities outside of academia. Such a course should be designed as a modular, credit-bearing elective, incorporating intensive workshops, personalized mentoring, and practical job exploration tasks. The workshops will address crucial subjects such as career diversity and skills transferability, while mentoring sessions will concentrate on individual career guidance and curriculum vitae (CV) development. The elective nature of the course allows for gradual integration into the academic program, helping to alleviate potential resistance from traditional academic circles by showcasing its value through voluntary engagement and positive results.

Dispelling misconceptions surrounding non-academic career paths involves challenging the misconception that these avenues are inferior or less

© The Author(s), under exclusive license to Springer Nature Switzerland AG 2024

S. Gaburro, E. Rieux, *Career Options in the Life Sciences: Guide Research for Your Path Beyond Academia*, https://doi.org/10.1007/978-3-031-73565-3_7

appealing. The course should emphasize the range of successful career shifts across various fields, from the humanities to the life sciences, illustrating how academic competencies are highly valued in sectors like technology, pharmaceuticals, and public policy. This diverse representation serves to rectify the inaccurate belief that academic skills hold little value in industrial environments. Moreover, involving non-tenured academics and postdoctoral researchers can provide a realistic perspective on the competitive nature of the academic job market and the challenges of securing tenure-track positions, underlining the need to prepare for diverse career trajectories early in one's educational pursuit.

Regular updates based on participant feedback, labor market dynamics, and continuous assessments are imperative to maintaining the course's relevance and efficacy. Building connections with industry stakeholders through guest lectures, case studies, and networking events enables students to witness the practical application of their skills and interact directly with potential employers. This enhances the course's practical significance and positions it as a vital component of a contemporary educational institution that aligns with global employment patterns and equips students for a prosperous career within and outside academia.

While the outlined course provides a strong foundation for expanding career opportunities for graduate students, it encounters limitations and obstacles that call for future improvements. The main challenge lies in institutional resistance to straying from conventional academic trajectories. Overcoming this obstacle typically necessitates backing from key university figures like deans or senior officials, who can advocate for the initiative and facilitate its integration within the academic framework. Demonstrating the course's value through data on labor market trends and alums success stories can aid in garnering this crucial support, for example, the Odyssey Program at the University of Cork's Odyssey Report 2022 | University College Cork [1].

In the foreseeable future, a noteworthy opportunity exists to enrich the course by establishing more robust industry partnerships. This endeavor may encompass the development of specialized modules tailored to directly address specific industry requirements, such as those in data science or technology. Implementing such modules could facilitate a matching process wherein companies actively engage in the educational sphere by offering real-world projects and challenges, enabling students to apply their academic competencies practically. Notably, this collaborative effort not only enhances the overall learning experience of the students but also aids companies in identifying and recruiting individuals whose skills align closely with their needs, effectively bridging the gap between academic institutions and industry. By adopting this forward-thinking approach, the course is poised to retain its position at the vanguard of educational innovation, consistently adapting to meet the evolving needs of students and the global labor market.

To effectively assess the efficacy of integrating a career-focused course into master's and PhD programs, it is imperative to establish key performance indicators (KPIs). These indicators should be meticulously designed to evaluate the course's immediate impact on students and its enduring influence on their career trajectories. Presented below is an elaborate discourse outlining pertinent KPIs for such a course, complemented by a follow-up scheme spanning a biennial period:

The KPIs for the recently incorporated career-centric course should encompass immediate metrics like student enrollment figures, completion rates, and student satisfaction ratings derived from end-of-course surveys gauging content's pertinence and instruction quality. Furthermore, supplementary KPIs ought to concentrate on measuring the efficacy of the integration, including the number of industry partnerships forged and the proportion of course content directly contributed by industry stakeholders. In the long run, KPIs should monitor the career advancements of graduates' post-course completion, encompassing metrics such as employment rates in non-academic positions within 6 months of graduation, the variety of industries accessed by alums, and salary benchmarks relative to peers who did not partake in the course. To capture this data comprehensively, a structured follow-up process should be executed at intervals of 6 months spanning a two-year timeframe, involving surveys, interviews, and meticulous analysis of employment data. This thorough approach ensures a comprehensive evaluation of the course's impact on students' preparedness and accomplishments in diverse professional realms beyond academia.

Reference

1. Odyssey Report 2022 (University College Cork, 2022). https://www.ucc.ie/en/hr/research/theodysseyprogrammeucc/odysseyreport2022/

Correction to: Testimonials

Correction to: Chapter 6 in: S. Gaburro, E. Rieux, *Career Options in the Life Sciences: Guide Research for Your Path Beyond Academia*, https://doi.org/10.1007/978-3-031-73565-3_6

The original version of the book was inadvertently published with the incorrect placement of the figures. In addition, the titles of the testimonials and the figure legends have been updated for all the figures of this chapter.

The corrections have been carried out in the chapter and the updated chapter has been approved by the authors.

The updated version of this chapter can be found at https://doi.org/10.1007/978-3-031-73565-3_6

© The Author(s), under exclusive license to Springer Nature Switzerland AG 2024

S. Gaburro, E. Rieux, *Career Options in the Life Sciences: Guide Research for Your Path Beyond Academia*, https://doi.org/10.1007/978-3-031-73565-3_8

SPRINGER NATURE

GPSR Compliance

The European Union's (EU) General Product Safety Regulation (GPSR) is a set of rules that requires consumer products to be safe and our obligations to ensure this.

If you have any concerns about our products, you can contact us on ProductSafety@springernature.com

In case Publisher is established outside the EU, the EU authorized representative is:

Springer Nature Customer Service Center GmbH
Europaplatz 3
69115 Heidelberg, Germany

The manufacturer's authorised representative in the EU is Springer Nature Customer Service Centre GmbH, Europaplatz 3, 69115 Heidelberg, Germany. If you have any concerns regarding our products, please contact ProductSafety@springernature.com

Printed and bound by CPI Group (UK) Ltd, Croydon, CR0 4YY
30/09/2026
02225578-0001